THE
NIGHT
PARADE

"A gorgeous invocation of the magic-haunted spaces between lived experience and folkloric traditions, between the living and the dead, between memory and story. I loved *The Night Parade*."

—Kelly Link, bestselling author of
Get in Trouble: Stories

"Beautifully written and imaginative, *The Night Parade* takes speculative nonfiction to new heights. Jami Nakamura Lin is both poet and storyteller, mystic and philosopher, teaching us to see the world differently, to suspend our disbelief, using mythology to interrogate our notions of family, grief, fear, love, and belonging. There is no other book like this—it's truly a stunning and visionary work of art."

—Jaquira Díaz, award-winning author of
Ordinary Girls: A Memoir

"Genre-defying and deeply poetic, *The Night Parade* invites the pandemonium within the personal and mythic to a roundtable where ancestors and folkloric creatures transform grief, memory, and mental illness into the tangible. Impossible to put down, gut-wrenching, and magical. I cannot think of a writer who has written so personally while acknowledging ancestral and cultural grief with such grace and honesty. A crucial and groundbreaking entry for the literature of the Asian Diaspora and explorations of mental illness."

—Sequoia Nagamatsu, bestselling author of
How High We Go in the Dark

THE
NIGHT
PARADE

A SPECULATIVE MEMOIR

———

JAMI NAKAMURA LIN

Illustrations by
Cori Nakamura Lin

THE
NIGHT
PARADE

MARINER
BOOKS

New York Boston

For my mother, Donna Lin;
my sisters, Cori and Kristi Lin;
my grandparents, Tom and Pat Nakamura,
and Toshiko and the late Dennis Lin;

and in memory of my father,
Charlie Lin (1961–2018).
Always.

起

PART 1

KI

———

承

PART 2

SHŌ

———

転

結

PART I

KI

In the beginning[1]—

In the beginning I choose kishōtenketsu, the Japanese version of the four-part narrative structure that flows from Chinese poetry. I need something. I have too much story and not enough shape. I overflow my banks.

And here is a form I can fit in the palm of my hand.
Better yet: a form that does not require conflict. (When in my life have I not shied away from conflict?)

The first part of kishōtenketsu is ki.

"This opening verse takes history or persons as its subject, and using metaphors and associations thereof, a variety of developments begin to unfold," explains Yang Zhai, a Chinese poet and scholar of the Yuan dynasty.[2] All I have to do is set up the situation.

Kishōtenketsu. Four parts. Simple, I think, just follow along.

The problem is: nothing is ever as simple as I imagine.

RYŪJIN

—

The Dragon King who
controls the tides and lives in a palace
at the bottom of the sea.

THE DRAGON KING

———

I

Listen—

In the presence of a story—if the story is a good one—time collapses.

This is why I am always telling it.

———

Once upon a time, long, long ago, my people knew a fundamental truth: the sea was coming for them. It was not their enemy—as island folk, much of their livelihood depended on fishing—but it was not their friend. The ocean was honored with the kind of reverence that looks a lot like fear.

Then the nation of Japan was isolated from the world. Years earlier, they had seen glimpses of the West in the black ships that sailed toward China, belching a putrid smoke that foretold doom. Afraid, they shut up their ports one by one, and settled into two centuries of isolation, protected by powers fiercer than any fleet: the ocean and the gods.

Eventually the people began to fear more and believe less, and the nation that had once hid from others showed its face, opened its mouth, and began to swallow everything around it.

Around this time my ancestors left their islands and sailed across the sea. My mother's father's family sailed from Hiroshima to California around 1898; her mother's family sailed from Okinawa to O'ahu around 1905. At least they ended up close to the water. They could see it, smell it.

My father's family, on the other hand, left Taiwan in 1972, and settled in the landlocked Midwest, transplanting their roots to a place where the lakefront is called "the beach."

In the suburbs of Chicago, my father had to find water wherever he could. When I was a child, he'd ready his tackle box in darkness, light leaching in only as he drove to Twin Lakes, the closest body of water that contained the largemouth bass he coveted. After catching a few fish off the pier, he'd toss them right back in.

I'd wake to find him humming in the kitchen in his khaki Bass Pro hat, returning the Styrofoam tub of earthworms to the fridge next to his jar of chou dofu, washing the fish stink off his hands. I could not understand the attraction.

When I see a bonefish slide through the water, I get the shivers, he explained when I asked. I was surprised. My father primarily traded in humor or advice; I had never heard him wax poetic or get the shivers, from cold or fear or otherwise.

Then again, when I was a child, his body hadn't yet begun to shut down. Nor had my mind. In that kitchen, both of us were as intact as we would ever be.

II

Once upon a time there was no clear demarcation between gods and men, good and evil, heaven and earth, for the fabric separating the world had not yet stiffened. Then there lived off the coast of Japan a young man named Urashima Tarō.[1] He was by all accounts a kind and gentle man who fished like all his ancestors before him, who made only enough to feed his belly, and who loved the ocean as much as my own father.

While walking along the shoreline one day, Tarō found a turtle beached in the sand. Turtles were auspicious; diviners would place their hands on their shells and examine the cracks for portents,[2] looking for the answer to the question we have always wanted to know: *How can I be safe?*

Tarō helped the turtle back into the water. He wasn't looking for anything in return, but as soon as the turtle touched the waves it transformed, as animals in these stories often do, into the beautiful daughter of Ryūjin, the Dragon King. She invited Tarō to her father's palace at the bottom of the sea.

What luck, he thought. The Dragon King was the god who controlled the wind and the rain and who possessed the jewels that caused the tides. He could be mercurial, his benevolence and brutality two sides of the same shell. Remember the time he broke all the jellyfish's bones.[3] Remember the time he used his jewels to help Empress Jingū conquer Korea.[4] (The Dragon King sometimes lies in wait and other times goes and eats; the gods, it seems, love imperialism.)

The villagers looked upon the Dragon King with awe, the same way they viewed the sea itself. Living by the water, you never forgot what the sea can give—the abundant catches that fueled life in the fishing villages, good weather for voyages—and what it could take: the fishermen whose bodies the ocean kept like a secret in its deep and endless grave. The mothers warned their children of the drowned men who follow the funayūrei—the ghost ships.[5] Beware the dead and their grasping hands.

The people paid their respects. In Okinawa, the people knew that life began and ended in Nirai Kanai, the underwater realm of the gods. In Taiwan, they worshipped Mazu, the patroness of the seas. In Japan, they commemorated Susanoo-no-mikoto, who watched over the storms, and Ebisu, who watched over the fishermen, and in the tenth month, they celebrated the festival of the Dragon King.

For it is said that if you give the gods their due, they will give bounty right back to you.[6] Otherwise, how easily gods can become monsters. These are the stories that burn in your memory, that make the world readable, that keep you alive.

Now, though, our young fisherman Urashima Tarō did not have to worry about the wrath of the Dragon King, for he had the sea princess at his side. They dove into the water past the epipelagic zone, the twilight zone, and the midnight zone, until reaching the ocean floor, where they entered the Dragon King's wondrous palace, whose walls were built of coral and fish scales and whose floor was covered in rugs of silk and sealskin.[7]

In gratitude for saving his daughter, the Dragon King offered Tarō her hand in marriage. (Tarō learned what we all know: a king is not so scary when you have his favor.) When the fisherman accepted, he was showered with jewels and food in three long days of eating and dancing and festivity.

After his stomach was full beyond bursting, he remembered that his own parents must be worried. He had now been gone for several days.

The princess gave him permission to return to land for one night only. Before he left, she gave him a black lacquer box. Take this, she said, and whatever happens, do not open it. He promised.

When he arrived on shore, nothing was as he remembered. Look at the pile of scrap where his house had once stood, smell the putrid funk of decay. And his parents—where were his parents?

He asked a passing child what had happened to the village. She looked at him in bafflement. Who are you? she asked.

I'm Urashima Tarō, said Urashima Tarō.

Urashima Tarō? the girl repeated. That's a name from legend. The man who disappeared.

She led him to a graveyard, the only part of the village that remained the same, where on a moss-covered stone he saw his parents' names. Here his crumpling knees, here his tears. One day in the Dragon King's palace down below is the equivalent of one hundred years up above. He did not know.

After he gathered himself, he thought he might as well return to the sea. There was nothing left for him here.

But the Dragon King would no longer have him. He had chosen the land, and now the sea said no. In despair, Tarō opened the box he had promised never to open. It released a puff of smoke. When it touched his face, he turned into an old man.

And so Urashima Tarō lived out the rest of his days on land, a man among phantoms.

When you hear this story, do you think it is better to be the parents, dead, or Urashima Tarō, alive, lost in a time no longer his own?

III

Here's another story. Once upon a time my father took me on a fishing trip to a tiny island off the coast of Belize. I was twenty-two, in my final spring break, vacillating between depression and anxiety over a future whose shape I could not see.

(When you hear *once upon a time*, you understand what kind of story you are hearing. A folktale. A fairy tale. When you hear *once upon a time*, you know to suspend your disbelief.)

This was my first fishing trip with my father, though he had taken both of my sisters before.

I like taking you girls on individual vacations, he told me, but people always stare. They think I have a very young wife. (And it was true that at the hotel on our stopover, the concierge had asked my father and me how many beds we wanted.)

All week it rained on that island, a glorified sandbar containing three huts and six people. My father fished and I read and wrote, and we both soaked in the peace that comes with isolation.

This trip was the longest my father and I had ever spent alone together. As we ate the jacks he caught with fry bread and watched the sea, I felt no particular compulsion to instigate deep conversation. Back then I thought time stretched out in front of us, limitless.

(In a year I would be hospitalized again; in seven years, he would be dead.)

On our return trip to the main island, a storm struck our three-person boat, pummeling us into one wave and then the next. Despite our captain's efforts we were knocked back, and knocked back. We had not appeased the Dragon King.

I cannot accurately describe the sheets of water or the motion—it was as if we would be swallowed. I only knew two things: one, that we would die; and two, that this thought was irrational.

I stoked both thoughts equally in my heart until my father grasped my slick hand and said *Lord save us* over and over.

It was then that fear gutted me. My father was my reality tester. *Am I sick? Is this okay?* I'd always ask him. *Is this true?* Now his face was scrunched shut, barraged with water.

So if he thought we were going to die, then we were going to die. I knew what the ocean could do. When I was a child my father's own cousin had disappeared on a diving expedition off the coast of Taiwan, his body never found.

Our boat rolled in the swells while my father prayed. Though I'd spent much of my life wooing death, I had not imagined it could happen like this, without my consent. Oh Lord, my father said, oh Lord, our fingers interlacing until later—how much time later, I have no idea— the bow hit the shore.

My father had loved fishing on that little island. He never went back again.

Over the years I've recounted this event many times, modulated for high drama or laughs or feigned nonchalance—*this weird thing happened.* Each time you tell a story, you can massage the tale to fit your needs. You can gauge the audience's reactions, alter the form or the tense or the point of view. With a little maneuvering, a little emphasis here and a little de-emphasis there, you can make an ending seem happier.

But not always.

Six years after our trip to Belize I received the call that my father was dying. I was twenty-eight, recently married—too young for such news, I thought, though I did not know how old was old enough. I walked the streets soaked in my own tears, imagining every possible way his death could take place, every kind of way it could unfold.

Some people, some Christians, expected my father to heal. They possessed a certainty I did not. They hearkened to the power of God.

I believed in such power—*the was, the is, the is to come*—I just also believed in his ability to withhold it.

The Dragon King gives and the Dragon King takes. With the flow-jewel he could make the waters rush in, with the ebb-jewel, he could make the waters recede. The tumors turned my father's skin first dry like rice paper, then shriveled and shrunken like nori. Though my mother labored over him daily with her lotion and her love, only in the bathtub did my father's itching body have a moment of relief.

When I was with my father, he was alive. When we sat around the fire, he was there and I was there and it was now. When I was not physically with him, I saw him transforming into something that once was, a ghost of a living man.

I do not want my father as myth. I want my father as father.

IV

Let's begin again, for the closer we are to the beginning, the farther we are from the end.

Once upon a time my father threw his own ashes into the sea. This sounds like a story but it's true. On the very last fishing trip before he died, he told my sister Kristi: It takes a long time to get a permit to scatter your ashes here in Hawai'i, so I'm just gonna do it now.

He opened a little tin that held his own clipped fingernails and flakes of dried skin. He sprinkled them into the water.

The ashes of my father fell down, down, down. Time on land passes differently than time under the sea. The sea embraces its fixedlessness.

By the time the ashes floated into the palace of the Dragon King at the bottom of the ocean, my father was dead. On land three weeks passed between when my father stood there in Hawai'i and when his body lay silent in our home. But in the abyssal zone, twenty thousand feet deep, it is always dark. How can you tell one morning from the next? The natives of this zone—the hagfish, the vampire squid—had to invent their own ways of marking time, of counting days till the end.

Urashima Tarō was a fisherman, not a native of the ocean floor. Without sunlight as a guide, time washed around him like a current.

———

My father would not have liked the story of Urashima Tarō. He loved a classic Hollywood story with clear conflict and a tidy conclusion and a plot that sucked you in immediately. He would sometimes turn off a TV show after fifteen minutes, saying, *It did not capture my imagination.* He did not like a sad or ambiguous ending, though in his own stories, we couldn't always tell if we were heading toward a moral or a punch line. Was that a joke? my sister Cori once asked him, after he made a comment out of the side of his mouth. He paused for a moment, then responded: Was it funny?

In Urashima Tarō's tale, there is no valiant warrior who slays the Dragon King, who in this tale is more absent than evil. There is no triumphant hero's return.

These kinds of folktales are hard for me to remember because their plot structures wobble in the middle, fracture, do not go where I expect them to go. They never reach the climax I anticipate. They do not provide the catharsis I want, the happy ending wrapped in a bow. I want Hansel to trick the witch into the oven. I want the woodcutter to kill the wolf.

But in this tale that has no villain, there is only a young man living out his days in a world to which he no longer belongs.

V

Tell me a story, my three-year-old says every time I bathe her. Except it is 2022, and I have no more stories. The ones I invent are milquetoast and meandering, peppered with characters from her favorite TV shows. The End, I say as soon as possible. She can always sniff out a false conclusion. What happened to *X*, she says. I want a *longer* story, she says.

When she is clean, she yells *Papa*, and my husband bounds into the bathroom, lifts her water-wrinkled body into her pink hooded towel, and swings her around in the air as she giggles.

Afterward I add more hot water for my own bath, dip my foot to check the temperature.

Be careful, J. Don't step on their legs, my father says. For here are my two little sisters, crowding the tub, and here is my father, sitting on the lid of the toilet, reading the *Chicago Tribune*. We are all so young. Look at his tinted aviators, his clavicle as sharp as his scalpel.

He washes our hair, our tiny bodies. We pour water on each other's heads. He lifts us up, for in this story, and this story only, my sisters and I are one. (Remember: *Once upon a time*. Remember what kind of story this is.)

He wraps us in our huge giraffe towel, our face covered by folds of fabric. We're going to Disney World, he whispers into the terry cloth, lifting the cocoon into a fireman's carry. We have to catch the plane. He holds us horizontally and zooms us to Florida.

We're climbing up the stairs to our hotel room, he says, lifting his knees high, jogging in our upstairs hallway. Here's Mickey! he says. Oh, you don't like Mickey? He runs away in a zigzag formation.

After a brief vacation, he says, time to fly back to O'Hare, my beauty, and this is the worst part of the trip—when we have to go home.

We draw it out as long as we can. In the cocoon it smells like dryer and wet and our own breath, and even after we are old enough to peel back the towel to see our upstairs hallway, we do not want to. Me in my towel, my sisters in theirs, my daughter in hers—our fathers lifting us—there we are, children suspended.

Here I am, in the bathtub, a grown woman with her face underwater, lurching for air.

———

In the presence of a story—if the story is a good one—time collapses.

This is why I am always telling it.

But the reverberations of a collapse spread outward. Shock after shock after shock. Sometimes, when my husband carries our daughter after her bath, swings her in the air, I close my eyes. Sometimes I turn away from the sea.

Neither the Dragon King nor the turtle shell of divination tell me the thing I most want to know, which is: How do you escape from a collapse unscathed? How do you avoid being buried?

In this story, the only thing I know is terror. And that terror is a god.

KAPPA

Water-loving yōkai who enjoy cucumbers and sumo
wrestling and are known for attacking humans in rivers.

AUTOMYTHOLOGIES I

———

The problem is that the story of my father and me is also the story of my illness and me, in the sense that all my stories spiral around illness. How sick I am of this.

Each time I open my archive—the hardback smiley-face journals from elementary school, the notebooks whose covers I collaged in high school, the staid black Moleskines from college—I am searching for the demarcation between Before and After.

Colloquially it is called a breakdown. But the term *mental breakdown* sounds like a singular event, as if your brain snaps as easily as fingers, or twigs.

A clean break, my doctor father once told me, *is the easiest kind to repair.*

The other kinds of fractures have crooked edges. Shards. Fragments that must be pinned to fit.

———

In May 2017, I woke in a youth hostel in Tōno, a little village tucked in the basin of a mountain range in the northeast prefecture of Iwate. I was twenty-eight, on a four-month writing fellowship to Japan. In less than two months I'd discover my father was dying; for now—blessed ignorance. I could look at the men passing in the street without a clutch in my chest. When I woke, I felt excitement, not the cold shock of remembrance.

The hostel's sole guest, I waited for my breakfast in a long wooden room, looking through the enormous windows at the rain falling on the even more enormous mountains, astonished at the scale of the landscape. I was from Chicago; I had never even used a parking brake.

Minutes later the proprietor, a middle-aged woman wearing a lacy pink apron, appeared with a heaping tray. A salmon wedge, a cup of soup, a glistening egg with neon yolk. Yogurt spooned with berry jam, a toasted slice of shokupan. Seaweed salad in a little bowl. I looked at each tiny treasure and wanted to cry.

Back in the cramped Tokyo studio that served as my home base, my breakfast consisted of whichever of yesterday's leftovers could fit in my refrigerator, an appliance so short that when I purchased a liter of milk, I had to decant it into two water bottles.

As she watched me eat, pleased at my delight, the woman asked about my plans for the day. Luckily, this was a question I understood in Japanese—a rare occurrence—though as I tried to align the verbs and nouns in my mouth to respond, I found I could not.

Kappa, I said finally, giving up. It was a word I'd only seen in print, and despite the way I'd practiced my vowels at home, I still pronounced it like the Greek letter.

She looked at me in confusion. I attempted the word again, shortening my long, nasally Chicago A's.

Oh! she said, after another attempt. Kappa!

In her mouth, the first syllable sounded like *cop*, the *p* popping like a little bubble. I nodded, flooded first with relief, then embarrassment. She pointed at a shelf on the wall, where sat a large green stuffed animal with wide eyes, a yellow beak, and dark petals atop its head—a toy that would fit right alongside Badtz-Maru and Keroppi. Looking at its smile, you would never guess that this creature was once best known for invading human's anuses to retrieve the ball of flesh that held their souls.

After I explained in simple nouns the places I planned to visit, the woman left me alone. I savored the flavors and textures of my meal, for here I could eat as slowly as I wanted. In Japan I was funded by grant money. I was beholden to no one. I hadn't had a bipolar episode in five years. It looked, and felt, a lot like freedom.

———

The village of Tōno is famous for being the cradle of Japanese folklore, the place where Yanagita Kunio, the man called the father of folklore studies, meticulously cataloged local lore into the 1910 compendium *Tōno monogatari*, or *The Legends of Tōno*. I was in Japan to write a young adult fantasy novel set in a world based on such stories, and so to Tōno I went.

Several of the tales in Yanagita's book concern the kappa, one of the most famous of all yōkai, the Japanese creatures and monsters and spirits of legend.

Yōkai are the frightening figures I'd encountered as a child in my brightly colored *Japanese Children's Favorite Stories* picture book, though there the word *yōkai* was never used. I didn't understand yōkai as a category until after graduate school, when—sick of myself, sick of writing about myself—I began to dive into these tales.

Despite my obsession with yōkai, I had trouble explaining to people what they were.

Yōkai are what Usagi Yojimbo fights, I said to my sisters and cousins, who like me had grown up reading Stan Sakai's manga about a ronin rabbit.

Yōkai are like the oni from Momotarō, I said to Japanese American friends. Think of *Spirited Away*. Think of Totoro and Pokémon.

Yōkai are like the equivalent of orcs and Ents and the Loch Ness monster and Big Foot and demons and ghosts, I said to everyone else.

I knew this was a pale comparison. I still struggle to form a brief, coherent answer. The kanji that make up the word *yōkai* refer to strange apparitions; they can be understood as beings that occur as a result of a supernatural event. Yet every book I've read on yōkai also touches on the slipperiness of its definition. As yōkai scholar Komatsu Kazuhiko cautions: "Yōkai has a broad range of meanings and these meanings are not fixed."[1]

To say they are spirits is to ignore all the yōkai who are defined by their visceral flesh. To say they are monsters is to flatten hundreds of different characters into one, to overlook all their individual personality traits, their mischief and humor and honor—and yes, their anger and violence and thirst for revenge. (Many kinds of yōkai *could* easily be categorized as monsters, and many could not.)

Yōkai, more broadly, are the uncanny, the answer to the question *Why is this happening?*[2] Yōkai are a way to make the formless concrete, to try to give a name to the nameless thing that keeps us up at night.

Yōkai are dynamic, transforming over the centuries as people discover or invent new ones, as other yōkai die out or disappear. And though yōkai can be playful—appearing frequently in comic forms, in manga and its precursors, in games—they reflect a culture's questions and fears.

"Yokai begin where language ends," writes yōkai scholar Michael Dylan Foster.[3] And yet here I am, trying to explain, mouth gaping open.

———

By midmorning Tōno's rain had dissipated, leaving only mist rolling down the mountains. I hopped on a little red pedal-assisted electric bicycle that I borrowed from the hostel. On the website I'd read the word *electric* and thought it would be like a moped, useful for getting me up and down miles of Tōno's hilly terrain. I imagined a vehicle that would require courage, or at least a thick pair of pants.

Instead, the electric bike looked exactly like a regular bike. When I pressed the electric-assist lever on the handle, it felt like a push from a not very strong child.

Help, I thought, pedaling as hard as I could up the inclined roads. Though I possessed what my ama—my father's mother—called daikon ashi, my legs were stuffed with fluff, not muscle. I stopped frequently to rest, gazing at the flooded rice paddies whose surface, unmarred by ripples, reflected the mountains like panes of glass.

That was the thing: I always needed to rest. Even here—a place where I did not have to report to anyone and all my notebooks were fresh and blank—I knew how travel could tip a person over. Here I was separated from my husband, family, therapist, psychiatrist, and regular schedule, and I couldn't even take my regular medications. In lieu of Adderall, which is illegal in Japan, my psychiatrist had to prescribe a less effective stimulant.* Neither my insurance nor my Chicago pharmacy would allow me more than a thirty-day supply of the controlled substance, and Japan wouldn't allow pharmaceuticals in the mail.

When I tallied the time spent trying to acquire this medication before my departure—hours spent researching, or on the phone with the consulate, multiple pharmacies, my insurance company, and the Japanese version of TSA—I came up with almost a full workweek, for which I received nothing except another Blue Cross Blue Shield

---

* In December 2021, after almost ten years of taking Adderall for fatigue and lack of focus, I was diagnosed with ADHD (in addition to my previously known bipolar diagnosis). But that is a different story.

customer service rep telling me *Sorry, the agent you spoke to earlier was wrong.*

At the last minute, my father learned from another doctor a loophole that would allow me two months' worth of pills. When I first received his voicemail laying out his plan, I was upset that he had done this without consulting me. By the time I called him back, half an hour later, I was grateful. (And now, years later, due to the vagaries of phone syncing, this is the last voice mail I have from him.)

Still, the sixty-day supply did not cover the entire four-month fellowship. In Japan I managed my intake, deciding which days were going to be Big Days, worthy of a pill, and when I would have to droop through a morning of writing into an afternoon nap.

The days in Tōno counted as Big Days. I pedaled, sweating the entire way, to Denshō-en Park, where images of the kappas' smiling faces greeted me from every corner, and where I paid two hundred yen for a kappa fishing permit. After being handed a kappa sun hat and a fishing pole baited with cucumber—the kappa's favorite food—I followed a stream to a spot called Kappabuchi, one of their favorite hangouts.

The kanji in the word *kappa* mean "water child," an apt name for a river-haunting creature the size of a three-year-old (or six-year-old or ten-year-old, depending on who you ask). And who you ask is important. To understand the story of the kappa you have to travel backward.

The characteristics that bridge visual depictions of kappa across time, from present day to the medieval scrolls, are the carapace on its back and the shallow indent on the top of its head, which contains its life energy. If this liquid spills, the kappa will weaken or die. While very strong, kappa are also unfailingly polite. When one is bowed to, it will usually bow back—thereby spilling its life force. This is good to know if you attempt to conquer one in a sumo match.

The contemporary kappa—the image that greets visitors in the gift shops and on the hostel owner's shelf—is cute and cuddly, but if you go back in time, its image is much more frightening. It looks like a gaunt hybrid of a monkey and a frog. The first textual mention of kappa appeared in the fifteenth century, and by the seventeenth century they were believed to be "aged soft-shell turtles that took on humanoid form."[4] The older stories emphasize the malevolence that accompanies kappa's mischief. It might drown your horse. It might drown you and eat your liver. In this way, the stories of kappa were protective: children knew to stay away from the rivers where kappa might live.

Yet the kappa is also capable of bestowing blessings. Around the Kappabuchi stream I found a child-size shrine where visitors had placed ema, the wooden plaques available at shrines and temples, upon which worshippers can write prayers. They'd also left offerings of stuffed animal kappa and a variety of breast-shaped handicrafts: tan breasts with dark crocheted nipples and pink fabric breasts with nipples of red buttons. Mothers who left tokens here, I learned, would be granted plentiful breast milk. (I had no child and did not leave anything. A year and a half later, staring at my low milk supply, I regretted this decision.)

I dropped my fishing pole into the water and waited. The cucumber bobbed in the water. But my father was the fisherman in the family, and I didn't know his secret ways. The kappa knew I was an interloper and did not bite.

In a photo from this day, I look thrilled, squatting in approximation of a kappa crouch, holding one hand up in a broad wave. In the other hand I grasp a fishing line, the thread I saw as explicitly connecting me with the past.

———

I love how *The Legends of Tōno* skirts the edges of my known world. The collection, published in 1910, firmly situates the kappa among humans. Yanagita relied on a twenty-five-year-old Tōno local named Sasaki Kizen for the stories, and Sasaki relied on his elders. One tale goes:

> When she was young, Sasaki Kizen's great-grandmother was playing with friends in the garden when she saw a boy with a dark red face behind three walnut trees. It was a kappa. Those big walnut trees are still there. The area around the house is now filled with walnut trees.[5]

The year it was published is a year I have touched. In 1910 my great-grandmother—my mother's father's mother—was a twenty-one-year-old living in Japan.

I met this great-grandmother once, in 1992, when I was three and she was one hundred and three. She would die shortly after this visit, my last direct ancestor who had ever lived in Japan.

Say hello, my mother said. Hello, I said grimly. The halls of the nursing home smelled of antiseptic and diaper cream and my great-grandmother herself was a small wrinkle lying in a sea of sheets. I asked to go outside.

While my mother and grandparents continued their visit, my father and I sat in the van with the sliding door open, eating stacks of turkey circles and butter crackers and sweaty squares of cheese. By the time I finished the dessert—one melty Andes mint whose dregs I sucked off the foil—it was time to go home.

The fact that I can't remember anything about my great-grandma yet can describe the specific food in every section of my Lunchables package accurately reflects both my three-year-old interests and my memory in general.

———

As I biked through Tōno's green expanse I worried about how I would later capture what I was seeing. I struggle with writing about physical space, for the same reason that in Tōno I often lost my way: a lack of spatial reasoning, combined with a faulty mind's eye.

I have a hard time understanding where my body is in relation to anything else, and I can't easily recall faces. At the library where I worked before leaving for the fellowship, I could speak to a patron for twenty minutes and still not recognize them the next day. When I went through a battery of psychological testing as a seventeen-year-old, the psychologist timed me as I tried to put together a three-dimensional wooden puzzle, a boat shape halved in a complicated way.

I can't do it, I said, after five minutes of twisting the pieces, trying to force them together like a baby with two blocks.

You have to keep trying, he said. I'll keep timing you until you get it. I tried. I pivoted. I rotated. I wondered how long he would let me struggle. The pieces would not fit.

My frustration and embarrassment spiked as time went on. This test was supposed to show what was wrong with me. I wanted it to illuminate the shadowy part of my brain and deem it *exceeds expectations*, but as the man watched me blankly, I knew that it would not.

All my life I'd tried to compensate for these mental deficits by taking copious notes and photographs. The inability to imagine landscapes and worlds was the whole reason I was in Japan; I'd applied for the fellowship explaining that research could only get me so far. *I need to put my feet in the dirt*, I explained. Without seeing a place with my own eyes, I could not evoke it in my writing.

I have no sense of what any of these characters or settings look like, a friend once said of my stories. Well, at least I have a sense of the *emotional* landscape, said my sister Cori, who always knew how to cushion a blow.

Taking a break from cycling, I looked at the Tōno basin and the mountains surrounding me and wrote in my journal: *It's as if I'm at the bottom of a bowl—the very last udon noodle, ready to be slurped*—which, years later, does not re-create much.

————

In Tōno, unlike the rest of Japan, kappa can have both red and green faces; both kinds appeared in the shop windows and restaurant displays. The illustrations, however, did not depict the kappa that *The Legends of Tōno* described: "In a house by the river . . . women have become pregnant with a kappa's children for two generations. When the kappa children are born they are hacked into pieces, put into small wine casks, and buried in the ground. They are grotesque."[6] (At what point, I wondered, did those women know they were pregnant with kappa? Was it only when they saw the hairy heads and bright red faces emerge from their wombs? Or was there some earlier indication that something in their body was not quite right?)

————

In the ninth and tenth centuries some eminent scholars, possessed of rarified knowledge, were known to have a special relationship with yōkai. Such scholars, it was known, could also resurrect the dead.[7]

Yet by and large, for many generations, yōkai tales like those of the kappa children endured through oral transmission—from grandparents to grandchildren, or through social games like the Hyakumonogatari Kaidankai, in which a group would light one hundred candlewicks, and each player would take a turn telling a frightening story and blowing out one wick until all the light was snuffed out.[8]

During the Edo period, these popular stories began to appear in printed formats, and by the end of the nineteenth century, they'd captured the interest of academics and the urban literati on a large scale.

Yet the first yōkai scholar of the modern age dedicated his work to erasing yōkai. As a proponent of Western scientific thought, philosopher Inoue Enryō believed all yōkai could be explained (*explained away*) through existing or forthcoming knowledge.[9]

Yōkai, he thought, represented a type of thinking both backward and backwoods. There he is with a taxonomy and a red brush, gleefully x-ing out monsters as modern explanations—static electricity, magnetic fields, wind patterns—killing off yōkai one by one.

But while a creature can be killed—or exorcised[10]—by a single person, the idea of a creature cannot. As long as there are things we fear and do not understand and for as long as we are still trying to understand them, there will be yōkai.

———

Yanagita Kunio resisted Inoue's approach. He argued for acceptance of ambiguity, believing that yōkai were relevant because they had historically played (and still currently play) an important cultural role. Whether or not they exist was somewhat beside the point.

He was more concerned with weeding out what he considered genuine tales—ones that had a historical footprint, a lineage in the oral tradition of a local culture—from ones that contemporary people had invented whole cloth.[11]

You could say—I could say—that what makes a story true is not whether it is literally based in fact, but whether at some point, someone believed that it was true.

As I looked through Yanagita's papers cataloged at the Tōno Folktale Museum, I wondered: How do you sift through all these tangles of threads to find the ones that accurately tell the story of an entire people? For that matter, how do you tell the story of a single person?

———

You could say that this story begins when I was eighteen, the year I formally received my diagnosis. Or you could say that the diagnosis was simply wrapping paper, a rubber-stamping of something that began long, long ago.

Lacking a memory, I try to follow my childhood through an extensive paper trail. My notebooks started at age five and expanded obsessively at age nine, filling three large bookshelves plus a stack of orange milk crates and banker's boxes. Then there is my virtual archive— my LiveJournal is 94,282 words; my Xanga is 35,040—the slush of documents on various hard drives.

People have spoken of my journaling as compulsive. Certainly, at times, my mania extended to graphomania; once, as a teenager, I filled every page in a journal in less than a week. From the time I was a child I worried over who would take care of my papers when I died. I figured if I became a famous enough writer, someone—a university?—would want them.

I don't know much, but I know the story begins before the doctors wrote the word *bipolar* next to my name.

———

Did it begin in 2006, when I was seventeen, the year of my psychological testing, the year of my first hospitalization? When I wrote: *My parents have determined I am having a mental breakdown . . . I would have to agree.*

Or earlier, when I was a freshman in high school, and chose to research bipolar disorder for my school project, though I wouldn't receive the diagnosis for another four years? When I wrote: *I have two minds, one tells the other what I should do. I listen to this other mind too much, it makes me feel separated from myself.*

Or even earlier, when I was twelve and started carrying four pink Benadryl to junior high every day to quell my racing thoughts? When

I wrote: *I will kill them with cyanide and then I will chop them up and pulverize them into a pulp and mash them up.* When I advocated for mass suicide in a school assignment, so the world would know that we students were too stressed? When I drew a graveyard with the caption: *Here lies Jami, who died of too much homework.*

She's always had a dark sense of humor, said my mother in the principal's office.

These diatribes peaked around my thirteenth birthday, as if whoever tossed me over the wall of adolescence did not throw me far enough and I broke open at the seam. In an expletive-laden entry written in purple marker, I write, *I hate them. I am really going to commit suicide. I hate them.*

The seventh-grade guidance counselor recommended therapy. (Perhaps this is where I begin: the point at which other people began to be afraid of me.)

The therapist said I was depressed. I'm her father, my father said, and I don't think she is depressed.

Do you want to keep going to therapy? my parents asked me.

I did not like therapy and each time I cried so hard I could not eat dinner after. No, I said.

Are you depressed? my parents asked me.

At thirteen, I already knew the truth.

To my parents, I said: No. Of course not. No.

———

Two points make a line, yet I still have trouble getting from there to here. My journal operates in jump cuts when I want a slow pan. Everything else is lost in the margins.

In high school, I formed these entries into the beginnings of a novel. In college, I used them as the backbone for the fragmented essay that got me into graduate school. In grad school, I refashioned the entries into scenes, filling in the gaps with attempts at physical description and atmospheric detail. But I do not know how to show. I only know how to tell. At best these additions were stilted stage direction, at worst voluminous upholstery attempting to disguise a warped frame.

———

"A funny thing happens when you're always writing, when you begin to narrate your life as you live it," writes Larissa Pham in an essay where she accidentally lies, misremembering an event until she digs through her own archive. "The crisp, fat braid of the story begins to overtake the impressions of memory—and it's true that was partly why I kept a diary, to pull a sieve through the disorganized world."[12]

If my journal functioned as a salve while I was writing it, it functions as a source of disorganization now. Trying to piece together the sheaves of loose-leaf paper and the half-filled notebooks, I begin to hate these old journals, their physical and psychic weight.

I also find evidence that journaling might actually *contribute* to memory loss: A 2018 study shows that people who recorded an event (through photos, writing, or documenting on social media) later had worse recall of the event itself than people who did not record it at all. The researchers concluded that "using media may prevent people from remembering the very events they are attempting to preserve."[13]

Once, in a cable car suspended high above Sydney, my father asked me to stop writing in my notebook; he wanted me to look outside at the vista below. We're in Australia, he told me. You should experience it.

My mother defended me: That's how she experiences it.

However, after writing things down, I often have no memory of the experience at all.

———

When I look through my old journals, I am searching for the cordon sanitaire. Susan Sontag, in her oft-cited *Illness as Metaphor* (indeed, in narratives of illness, one cannot escape it), imagined a physical border between her kingdom of the sick versus the kingdom of the well, of which "everyone who is born holds dual citizenship . . . although we all prefer to use only the good passport, sooner or later each of us is obliged, at least for a spell, to identify ourselves as citizens of that other place."[14]

Yet mapping the terrain of my mental illness is futile. Any story I tell is a narrative choice, a question of aesthetics: where to aim the camera lens, and most important, what to leave unfocused.

"What is placed in or left out of the archive is a political act, dictated by the archivist and the political context in which she lives," writes Carmen Maria Machado in her genre-bending memoir *In the Dream House*, its own reckoning with an archive.[15]

The archive of my life is the archive of my ghosts. Reading it, you would think my entire life was haunted.

Every date I don't record: remember, I was happy then.

———

When researching yōkai I want to find the Ur-Tale, the origin story, behind each creature. I don't read Japanese, and didn't grow up in Japan, and do not know what attributes of each folktale were considered the main point—the equivalent of the wolf saying *the better to see you with*—and which details do not matter.

Folktales resist such efforts to designate an official canon. In the introduction to his *Guide to the Japanese Folktale*, directed at children, Yanagita writes about how the stories might deviate from the versions a child had heard before: "It does not mean that one or the other is a mistake or your memory is wrong . . . While the same story is being recalled over and over, it is only human that . . . those interesting parts are told in special detail and the rest are gradually overlooked or changed."[16] In Tōno, I took 434 photos over three days. I wanted to make sure I had enough documentation. I did not yet know what part of the story constituted the interesting parts.

———

We write to remember, but we also write to forget. My notebooks functioned as a place to store my thoughts so my brain did not have to contain them.

The problem with the journals is that I cannot find my way back. Many of my entries are brief, bullet-pointed, meant only to function as a trigger for the rest of the scene. Fifteen years later, reading them, I am often baffled. Who was V, the boy from MySpace? Who was Kylie, and why did I think she deserved to die? It is an argot I can no longer parse. The past, in this way, is unknowable to me.

The other problem is that I only look through the archive through the lens of my illness. I only seek portents. I ignore the large swaths of time where my mind was still, and my parents and I orbited each other in a gentle sort of stasis. This, too, was my life.

Is it a comfort that every historian, ethnographer, archivist faces such issues? (It is not.) Yanagita's version of folklore studies is, as contemporary yōkai scholar Michael Dylan Foster describes, "a discipline driven by nostalgic sentiment, informed by a desire to incorporate aspects of the past into the construction of life in the present."[17]

Though he is known as the father of Japanese folklore, Yanagita, a citified outsider, did not gather most of the tales in *The Legends of Tōno*. The collection is his literary version of the folktales collected by Sasaki Kizen. Sasaki was the one who lived among the people, whose great-grandmother's encounters with ghosts appear in the work, who had thought these stories important enough to remember. Yanagita, however, was the one who published them. The one who gets the credit.

I can only read the Japanese folklore scholars in translation, and many of their critical works were never translated into English at all. Sometimes I can only find fragments of arguments, embedded in criticism and commentary written by white male scholars over the past one hundred and fifty years. Other times, I simply am not looking in the right place. Lacking enough context or source material, I struggle to initiate my own conclusions, hamstrung by the links already presented.

Without my translator Lisa Hofmann-Kuroda, for example, I would not have understood Yanagita's role in prewar Japanese nationalism. (Folktales and fairy tales often play this role across cultures. The Brothers Grimm, who claimed their tales were "purely German in their origins,"[18] also contributed to rising nationalist sentiment.)

As Lisa writes, explaining the work of scholars Marilyn Ivy[19] and Gerald Figal:[20]

> Yanagita's project of collecting folktales and beliefs cannot be understood outside Japan's larger project of modern nation-building. This project required for legitimacy the invention of a narrative that implied the existence of a timeless Japanese "people" with certain unchanging characteristics . . . The project of "inventing" Japan was necessary to form a cohesive national identity as Japan extended its empire across the Asian continent.

Here, then, lies the seductiveness of folklore, and the ways it can be used to centralize power. "Yanagita . . . read[s] tales of tengu and other monsters as narrative embodiments of a collective Japanese psyche," writes Figal.[21] These are the stories that belong to us. This is who we are. This is our psyche.[22]

Here are your roots: now, you are rooted.

———

Sifting through my own archive, every arc I draw seems wrong. Only in other stories can I find my way back to something that seems true. I trace the lines of myths because the path backward is easier to follow than my own. The characters in the old tales do not emotionally resist. (To pick up remains and call them ruins—to pick up ruins and tell them: *Remain*—)

In Yanagita's theory of ambiguity I want to be able to disregard the *why*, the *when*, the *how*. Yet I know I will always keep flipping the pages. Contorting my body, my tales, into positions cramped and intolerable, as I seek the loose thread I can follow to the end.

Of course, there's that idiom about the definition of insanity (*doing the same thing over and over again, but expecting a different result*). The quote is usually attributed to Albert Einstein, giving it significant gravity. The phrase was really spoken by a character in a 1983 book by Rita Mae Brown, whose cozy mysteries (*Murder She Meowed, Santa Clawed, A Hiss Before Dying*) were popular with the patrons at the public library where I worked.

I want to find the blade that will cut my Gordian knot.

———

On my last full day in Tōno I biked six kilometers uphill from my hostel to the re-created historical village of Tōno Furusato Village, a long trek made longer by the many times I had to stop to rest my calves. When I arrived, I spent an hour measuring the dimensions of a magariya—a traditional thatched L-shaped house with a stable attached—using my feet. I visited villages like this all over Japan, trying to envision what nineteenth-century rural Japan looked like, a present interpretation based on a past interpretation of an even farther past.

Of course these re-created historical spaces develop because there was, first, a legitimate historical tie. The re-creation does not grow in a vacuum, which makes it harder to separate where the (heavily researched) fictionalization begins. In all my journals I was trying to present something, if only to myself.

Over generations the kappa figure changed from the one depicted in folktales—an ugly creature malevolent and mischievous by turns, who sometimes provided for the villages that gave it offerings—to the one depicted in folklorism, the bright green smiling cartoon character, yōkai scholar Foster argues. This change occurred when the kappa was ascribed value by people outside of the "folk group to whom it was indigenous."[23] The kappa of folklorism is the commodified kappa. It is sometimes difficult to ascertain when folklore becomes folklorism, but you can identify the trend lines—when the story becomes scrubbed and smooth.

Held at one remove, the kappa no longer terrifies. At one of many Tōno gift shops, I sifted through an enormous quantity of kappa-themed wear to find something to signify my trip; I settled on a pair of lime-green ankle socks.

And yet this kappa does not interest me. People adore this kappa the way we like all benign things. My youngest sister, Kristi, crocheted me an amigurumi kappa holding a sign that said *Gambatte!* ("You can do it! Do your best!"). I love it and display it on my shelf, but it emits no whiff of the ineffable. You can purchase the kappa of folklorism, you can hold it in your hand.

———

I did not have any bipolar episodes in Japan, despite my limited medication, or the fact that when I returned to my home base in Tokyo I sometimes didn't eat enough and stayed out too late and slept too little.

On my Tokyo weekends, the last train to my tiny station ran at 12:30 A.M. A couple times, after missing the train, I'd walk the fifty minutes home using my cell phone as a flashlight. I'd buy two ice cream bars from a vending machine to keep me company on the way. I was fine, but I could see how easy it would be to tip over into *not fine*. I could feel the border.

What no one told me, as my bipolar disorder stabilized over the years, was that sometimes I would feel an intermittent longing for the feral Before. That I would desire to cross that border and see the kappa of folklorism transform back into the kappa of folklore.

When you return to the humans you return to their laws. There are days I look at my life and think: *tamed*—the house, the child, the job. Sometimes I miss the frenzy of my adolescence, those kicks of adrenaline.

These are the moments when I forget what it was really like, mythologize it into a dark whirlwind of mysterious and creative force instead of what it was: depression and mania as two sides of hell's coin.

———

Stories, too, fall apart. Upon further investigation, I learn that mystery writer Rita Mae Brown probably did not come up with the line about insanity after all—two years before her book came out, an article in the *Knoxville News Sentinel* quoted an anonymous Al-Anon participant as saying the same thing.[24] When Sasaki Kizen, the Tōno local who had collected all the oral folktales, read the published version of *The Legends of Tōno*, he wrote to Yanagita: "The tales are not like anything that I remember telling you."[25]

Wherever we long to return to does not exist, if it ever did. Our present is ever disappearing in the future's gaze. You can sink under the weight of such knowledge, unless you periodically wipe the slate clean of modifiers that no longer have a referent. If there is a line between regurgitation and reincarnation, I do not yet know it, though I chide myself to let sleeping blogs lie.

At some point, during the telling of this story—between beginning this book and finishing it—I stop consulting the archive.

This is not a conscious decision. Instead, one day without looking at it becomes another, and then there it is, lurking in the corner, and with each day that passes its shadow gets bigger and bigger. Go in, I tell myself, and I cannot. I get too lost in its pages.

It sits and watches me. If I go in again, I don't know if I will ever get out.

———

On my last day in Tōno, I asked the proprietor of the hostel if I could stay longer. I wanted to return to the museum and the historical site and take more photos, take more notes. I wanted to make sense of it. By that I mean: I wanted it to make sense.

She paused, wiped her hands on her pink apron. She used the word *chotto*, meaning "a little . . ." meaning no. She apologized, but my rudimentary Japanese comprehension couldn't parse the reasoning.

I left the hostel on foot, dragging my little suitcase behind me. Each house was so far away from the next, with only rice paddies and low fields between them. I saw an old woman bent over a field with a red tiller, the same kind used a hundred years ago. I photographed her outline with the mountains rising in the back. I wanted to get closer and capture her face, what I imagined as a representation of the present and the past crumbling in on themselves. But it seemed too invasive.

I don't remember how I made my way back to Tokyo. Did I walk from the hostel to a bus stop? Did I walk to a taxi stand or all the way to the train station? I don't know. I did not write it down. In any case, I got home.

———

The kappa now has only one name, though it used to have many, each indigenous to its own region: kawaro, mizuchi, gataro, enko, gawappa.[26] The further back I go, the more layered its story becomes. I want it to bow to me; it will not bow.

You can write a story in a vacuum, but to tell a story requires a listener. When you tell a tale it is no longer yours. This is its purpose and its curse.

Of his *Guide to the Japanese Folktale*, Yanagita writes: "But under the severe circumstances of these times it is possible to think there are many whose souls seek peace, and we hope, therefore, that the publication of this volume will serve as a means of rehabilitating our people."[27]

When I first read this quote, I thought the key words were *peace, rehabilitate*. It seemed a tall ask of a folktale.* Of lives that will always have gaps.

But the key word, of course, is *seek*.

When I look through my archive, my boxes of journals, I am looking for an origin I will never find. I wonder, like philosopher Walter Benjamin, whether "each and every illness might not be curable if only it were carried along a river of stories all the way to the river's mouth."[28] But he did not leave directions on how to go upstream.

My hand keeps moving across the page. I revise and revise and revise, each time hoping that this version, finally, is a story I can live with.

---

* This quote is also ironic considering Yanagita's ties to nationalism and Japan's increasing imperialism when his scholarship was published.

PART 2

SHŌ

The second part of kishōtenketsu is shō: a continuation.[1]
In this second part of the four-part structure, you develop
the situations and characters set up in the first part.

Shō: a deepening. Just keep following the path
you were on. It seems easy enough.

"This second verse should be quiet and moderate," explains Yang
Zhai.[2] "It is important to avoid being too erratic or blatant in this
section, but by the same token, neither should one be boring."

You wonder: How not to be too erratic? To be neither
too wild nor too open nor too boring?

How to toe that line?

ONI-BABA

—

Formerly human women who have transformed into
demon hags and often feed on human flesh.

THE RAGE

Let me try again.

I

To be bipolar is to be hungry. By which I mean: to be bipolar is to deny your hunger.

But the same could be said about being a woman. About being a girl. From the beginning all the stories warned me about the consequences of unchecked appetite, of untamed desire. I swallowed their morals in one gulp. Eve and the apple. Persephone and the pomegranate. The woman from Mino and the village boys.

Of course, the woman from Mino has been forgotten by this world. I did not learn her story until I was a woman myself. This is no surprise, as this is what her story is about: being forgotten.

Once upon a time, almost a thousand years ago, there lived a woman and a man.[1] At first things between the lovers went swimmingly, as these things do, and then he began to avoid her, as these things go.

When she expressed her unhappiness he withdrew even further, until he stopped coming at all.

When she realized she'd been forgotten, the woman from Mino stopped eating. She waited for her family to notice her hollow cheeks, but they didn't. When she stopped getting out of bed, no one noticed that either.

Finally (as always, so much of the story happens in the gap before the *finally*), the woman from Mino used old millet jelly to slick her unwashed hair into five tall knots at the crown of her head. When the jelly dried, the knots looked like the horns of the oni, the demonic ogres* of legend. She admired herself in the mirror. She changed into pants the color of blood. She ran off into the night.

Dressed like an oni, she tracked down her ex-lover and killed him.

Here the story takes a turn, for then her horn-hair became actual horns. She developed a taste for child-flesh. She had transformed into a real oni-baba—a demon hag.

This story does not tell me the thing I most want to know: *how* the transformation from woman to oni-baba occurred. Did it happen all at once, as in Disney movies and fairy tales, or did the woman from Mino look at her reflection one day in the water and see the hint of something she did not recognize?

As a child, I read stories for the same reason I later wrote them: out of pleasure, out of curiosity, out of joy—and with the hope that one day, I'd find a story that would show me how to vanquish myself.

---

* Oni are sometimes translated as ogres or demons, but neither of these quite captures the term; see the chapter "In the Whirlpools."

II

Once there was a time when the word *rage* had nothing to do with me. I know this in the way I know that once I was a child. But those years exist more like a collection of photos without emotional texture.

Here's a snapshot: little Jami in a denim jumper and velvet shoes and a thick bowl cut that mushrooms around her face. At five and six I am obsessed with *Little House on the Prairie*, my baby doll Sara and my lamb Lambie, and the *Full House* reruns I watch every evening while my mother makes dinner.

Most of all I am obsessed with bouncing a kickball through the rooms of our open-plan house. As I bounce, I make up stories. The rhythm of the bounce and the movement of my body help the stories come out right, as if without my body my mind cannot function.

To the chagrin of Cori, with whom I share a room, my stories do not go to bed when we do. Long after our parents kiss our foreheads and shut the door, I'm telling stories into the pillow.

Can't you just say them in your head? she asks from the bunk below. No, I say, though I try to whisper. I can still hear you, she says.

This is our nightly back-and-forth, until enough practice helps me tell them silently.

My stories occur at the intersection of my two childhood states: wonder and frustration. I am a tentative child, forever afraid of getting in trouble. (Here I am at three, shamefaced, trying to clean up the pee dribbled down the side of the toilet. Here I am at four, hiding the burn on my hand from where I've accidentally touched my ama's iron. The untreated burn leaves a berry-size scar that mottles my hand to this day.)

I feel all of childhood's impotence and none of its freedoms. I dream of getting older; adults are the cause of every closed door. When I am a grown-up, I think, I will open them all.

Thus, my favorite game at this time is Donna-Tina, which I only play with my mother. I love the sound of the name when it runs quick off my tongue, like *sonatina*, and how in the world of the game, my mother and I are peers. We are grown-up women who do very grown-up things, like eat lunch at restaurants. In Donna-Tina I choose what we do and when we do it.

But even in Donna-Tina there are limits. I always have to be Donna, for instance, because my mother doesn't want me to call her by her real first name, and she is Tina, after my father's younger sister who babysits us often. But the point of the game isn't to pretend to be my mother or aunt, the point of the game is power.

One day, after playing, I tell my mother that I can't wait until I grow up and can call her Donna for real. She doesn't understand what I mean.

You know, I say, how grown-ups call their parents by their real names.

No, they don't, she says, confused. I still call my mother Mom.

I'd thought that once I turned eighteen, I would reach equal footing with my parents. Now I know that no matter how old I become, I will still be a child to someone. I am devastated by this news and mull over it for days.

———

My mother undergoes her own transformation on Wednesdays when my father works late. After serving my sisters and I dinner—often pizza, since the smell nauseates my father—she ducks under the dinner table. Upon reappearance, she's been replaced by someone whose eyebrows pinch and nostrils flare and whose steely eyes glare at each of us in turn.

My name, this person says in a warbly voice, is Miss Viola Swamp.

Upon speaking the name, our mother fully changes into a wizened woman with a warty nose and big black dress and black hair curled into two poofs on her head. My sisters and I shriek. We recognize Miss Viola Swamp from our book *Miss Nelson is Missing!* Miss Viola Swamp is the nice teacher Miss Nelson's doppelgänger, who subs when the students misbehave. She looks like a witch, or an oni-baba.

Miss Viola Swamp is frightening and delicious. Her cackle! Her jutting chin! The way she has my mother's body but not my mother's carriage or demeanor! We spend all of dinner asking Miss Viola Swamp questions, hearing her creaky, cranky answers.

When we have eaten our pizza all the way to the crusts, Miss Viola Swamp announces that it is time for her to go home. She disappears under the table, and the head that pops back up belongs to my mother.

How was dinner? my mother asks, all softness and innocence. How were you while I was gone?

I love our Miss Viola Swamp, who arrives with the dinner and disappears before dessert. She never appears when my mother does not want her.

But the character Miss Viola Swamp disturbs me. When my mother reads us the book, I get confused about the nature of reality. It's clear she is only Miss Nelson in a different outfit, but Miss Viola Swamp is alluring in a way Miss Nelson could never be. Miss Viola Swamp is the hot core, the narrative center—as if she is the real character, and the nice teacher the costume.

———

My mother and father love me very much, and they are, at this time in my life, strict in the way of conservative Christians. I am their first child, and, as such, a beta test. They are my parents and also—for much of elementary school—my teachers. My mother handles the bulk of the homeschooling, and my father teaches science and acts as our principal.

Every third time I forget to do my homework, the teacher calls the principal and leaves a message. After he finishes seeing a patient, my father returns my mother's call, and she hands me the phone for a talking-to. The fear of the call is enough. The fear, you could say, of God.

In every sector of my life, they are in charge.

When I am an adult, my mother apologizes for being too hard on a little girl, but as a child I think the wrongness is me. The only solution is to be very, very good.

And so I see childhood as an endless litany of tiny grievances: not being able to watch this movie or wear those jeans to church, or how my younger sisters have the same bedtime as me. It is easy to scoff at these petty things, but my world as a child is so small, and so in relative proportions, each No looms large.

Other people, other brains, can brush off these things. I cannot.

After my father, amused at one of my indignations, tells me youth is wasted on the young, I run to my journal and scrawl: *Adulthood is wasted on the old.*

It is hard to be very, very good when you are also prone to getting very, very angry. In my body something begins to build.

The problem is that I want you to understand how deeply I fear me without making you also fear me. This is an impossible tightrope.

For every story about mental illness, the unknowable churn of the mind, is also about the body, its fallible vessel. Once the Rage was in my body, I could not get it out.

III

It is easy—perhaps the easiest—to make the past into a tidy narrative. To say that I told stories because it helped me survive. But I also wonder if the stories and the Rage grew up together like spurious correlations, independent of each other.

By this I mean that stories were passed down to me from our family like other traits: thick eyebrows, big nostrils, clear skin. Twice a year my mother's family gathers at my grandparents' house in the East Bay, all the children and grandchildren and assorted aunts and uncles and second cousins sleeping on the floor, on the couch.

There the stories spread in every corner. In the backyard my father holds court in the hot tub, spinning tales to our cousins of us fighting supervillains who can turn invisible when no one's looking, or who can fart at will. (Only later will we discover the 1999 film *Mystery Men* and realize that all the adventures we starred in were riffs on his favorite sci-fi and action films.)

Afterward, when we're too wrinkly, my sisters and cousins act out these scenes on the grass next to the rosehip bushes and the aloe and the persimmon tree, whose silky fruit my grandfather mails to us every autumn.

I am three to six years older, and while I love listening to my father's stories, I do not want to play and pretend. Plus, I am a snob about age.

Instead, I bounce my ball, its movements soothing as I eavesdrop on the kitchen table, where my grandparents and aunties and great-aunties share stories around last night's leftovers or Costco pizza or YO! Sushi. At my grandparents' house food appears at all hours of the day like magic. Despite this, none of the women of my grandmother's generation seem to eat at all. These aunties are as thin as they were in their photos from the 1960s, with their black hair pouffed like Jackie Kennedy's.

Aren't *you* hungry? we ask, once we are old enough to notice. They always demur, and instead press upon us grandchildren another heaping plate.

Around that table I hear about California in the 1900s, and O'ahu in the 1920s, and Amache incarceration camp in the 1940s, and Chicago in the 1960s. I learn about how when my grandma was seven, they lived a few miles from Pearl Harbor, and when the bombs came her brother Howard—barely older than her—carried her on his back so they could escape.

I think that's why he grew up to be so short, she concludes.

From listening to these stories, I come to understand how a family mythology is built. I hear the peals of laughter. As I bounce my ball, I feel a sense of deep love and well-being. But I also learn what is shared, and what is not.

They do not speak of the mind, only of the body and its ailing ways. Only much later will I learn how bipolar appears on both my grandmother's and grandfather's sides of the family. Only much later will I learn about the kinds of trauma that some of the women at that table experienced. Around the table no one gets mad or irritated. Those stories do not appear.

To not let these stories out means we must swallow them secretly. For we know what happens to hungry girls. Look at the woman from Mino. In one moment, you're angry, in the next you're eating children.

———

In kindergarten, my mother splits my world like an avocado by teaching me how to read. It is not adulthood, but it is, I am sure, the next best thing. The girls whose stories I am drawn to, no matter who they are or where they are from (Ramona's Klickitat Street, Harriet the Spy's Upper East Side, Laura Ingalls Wilder's Little House), are in many ways the same girl. A girl who is curious and bright, whose arc involves her sense of injustice at the state of being a child. Of being a girl.

(It does not occur to me at the time that there could ever be such a book featuring an Asian American girl. Though my mother carefully curates our library to include Asian American authors and characters, the chapter books available are mostly located in the past. The Japanese American books concern incarceration, and the rest inevitably deal with some sort of immigrant struggle. And I want to read about spunk and angst, not pain.)

Long after I learn to read on my own, I ask my mother to read aloud to me while I lie in the patch of light atop our couch. I can still feel it: the nubby gray fabric pressing into my cheek, that square of sunlight, my mother's voice, the story soaking into my ears, a pleasure I cannot name.

As I age, such concerns continually appear in my own writing. From a story called "Florence's Trials," written when I was eleven: *I'm fifteen years old, for heaven sakes. I can make up my own mind,* and she quietly stole down and walked out of her house. She was a clever, romantic, dramatic girl.

In my stories I set the rules. I can be both god and hero.

It is nice, sometimes, to be the hero. By the time I am eleven, I know that I am also sometimes the monster.

I AM A BULL TODAY

February 28, 2000 (age eleven)

Steam rising from my ears
Smoke coming out of my nose
My arms crossed,
my fists clenched
my lips pursed,
the lower one jutting out
my shoulders are raised
my eyes bore a hole through your body
I hope your brain falls through it.

By now, the frustrations that every child feels—the tantrums that every child throws—have changed. I cannot articulate the transformation my body undergoes when my mind spirals, or the lack of control I possess during those times; I only understand that I am split in two. There is the part of me that looks and feels human, and the part of me that is wild.

At eleven I call this wild part *Bad Jami*; at twelve it is *my twin*; at fifteen, *my Hulk Self*. At seventeen, I name this emotional and physiological state *the Rage*. The Rage is not a yōkai that suddenly appeared; like rot, it accretes over time.

The Rage is the implosion that happens after an argument; after retreating into my bedroom my anger morphs into a head-to-toe fury. Breath and heart quick, bees in my limbs, hands clenching and unclenching, twisted fistfuls of hair.

The Rage is the part of me that fantasizes in my seventh-grade journal about murdering myself and others. It is the part of me that throws a book at my mother's face, that scratches diatribes into my doorframe with a knife, that requires handfuls of sleeping pills to sublimate. It is the part of me that hungers for destruction, and the part that I keep tightly locked within the confines of my bedroom in the suburbs of Chicago.

My public-facing self, the *Good Jami*, is always shocked by these thoughts and behaviors. Afterward, I privately apologize in my journal—to myself, to my parents, to God. I write, *Why am I like this, why won't the Rage go away, why won't my body stop shaking.* I do not know when or why a rage will strike; I only know that it will happen again. The Rage is immutable.

As an adult, I learn that another name for the Rage is *psychomotor agitation*, a condition especially pronounced in bipolar disorder and schizophrenia.[2] Symptoms include "sudden, unexplainable movements, inability to sit still . . . physical tremors, inability to relieve tension, anxiety, frustration . . . People with psychomotor agitation and mental health conditions such as bipolar disorder and schizophrenia may feel uneasy, nervous, or that there is no hope of controlling their agitations."[3]

While informative, the term *psychomotor agitation* lacks *the Rage*'s teeth.

IV

My anger contaminates everything around me. All those arguments: Me versus my parents. Me versus one sister, then the other, whose words I barrel over, using my own like blades. Me versus the whole family, our voices rising higher and higher over the dinner table or in the car. Me versus myself.

In junior high, I tell a couple different friends about my recent fights, and both tell me they would—could—never speak to their parents like that. Only then do I realize that perhaps the way I communicate is abnormal. The Rage is my own, a dark side no one else possesses.

But what other way is there to be mad? In the days before social media, before I know what community the Internet can hold, I see only what is in books and what is before me. I see no visible anger in my extended families, or in the Japanese American church that forms my family's primary community. We take care of each other, laugh, and grieve together, tsk at each other in secret. Any anger is tamped down, excluded from the warp and weft of our narrative fabric.

And I never see my own parents angry at each other. The anecdotes my mother tells me from their early years, before I was born—my father snapping at her about the lamb she'd cooked for dinner, her scraping it into the trash—seem from another world. By the time I am paying attention, the years have softened them, widened the margins of their love that expresses itself so visibly: in the hands they hold in the van, the way they dance together at every wedding. When I am especially angry, I write in my journal: *I wish Mom and Dad would get*

DIVORCED. I want there to be a wedge between them, a space where I can fit.

When my mother gets mad it stems from exhaustion and frustration, not fury. It does not seem related to what curdles in my own chest.

My father's anger *does* scare me, though I see it only maybe twice in my life. When I am ten or eleven, my own rage just emerging, my father drives us home from church. We are a couple blocks away, passing the abandoned lots and stalled housing developments on Old Plum Grove Road, and I am asking the same question repeatedly, trying to get the answer I want. He's getting increasingly frustrated. I keep asking. Finally, he screams my name, *Jamiiiiii IIIiiiii*, an undulating roar whose final vowel skates up and down three notes. The five of us freeze. It is the first time in my life I have heard my father like this.

Charlie, my mother says. He stops the car on the side of the road, shaking as hard as I am. My mother opens the side door of the van and gestures for me. My sisters in the back seat are still, eyes wide. My father grips the steering wheel and looks straight ahead.

My mother and I walk home, past the cul-de-sacs that are named alphabetically—Alder Court, Butterfield Court, Croftwood, Dahlia— past the empty green lawn we call the Big Field. We do not speak, even though I know she took me out of the car not as a punishment, but as a mercy.

Later that day he apologizes. His roar happens only one more time, six years later. He, too, was a swallower. Perhaps he feared his rage as much as I feared my own.

————

There are ways to defeat yōkai, ways to diminish their power. To defeat a kappa, you can trick it into bowing to you. To weaken an oni, you can take away its iron mallet.

There are ways to alleviate the Rage. Sometimes they are constructive. Sometimes they work. Imagine ten candles and blow them out one by one. Write in a journal. Squeeze a ball. Pray, if you have the wherewithal.

But those methods are not nearly as effective as the methods I come up with by myself. Swallow some pills. Slap paint on your wall. Bite your skin hard. Pinch your thighs with your fingernails. Etch curses into your door, into yourself.

One night in high school, unable to shake the Rage out of me, I take my purple glitter flip phone and bend the top half backward. The more I bend, the more the Rage lessens, until the hinge cracks and I have one piece of phone in each hand. I stare at them, startled, before hurling the pieces out my bedroom window into the tall bushes lining the backyard. My body cools as quickly as if plasma were injected into my veins.

An hour or so later, after the Rage passes, I tell my father. I wait for a lecture.

Where did you throw it? he asks. I point.

I'll go find the SIM card, he says. You'll need it for the new phone.

After he leaves the bedroom, I admonish myself. Again, I have lost control; again, my father is the one literally picking up the pieces. Every time I pray, *Let this be the last time.* Still, I do not offer to go find the pieces myself.

From my window I see him hunting through the bushes on his hands and knees, his flashlight a small beam of light in the dark backyard.

V

The only reason anyone remembered the woman from Mino was because of her insatiable hunger. Decades after her disappearance, long after her own family had died, the local villagers heard of a yōkai living in an abandoned temple.

This yōkai, they thought, must be the one capturing and eating their sons. They decided to set the temple on fire. As the flames rose, an oni-baba emerged from the building, wearing a tattered shred of red fabric.

Wait! she cried out. Before you kill me, let me tell my story first!

They looked at each other. She was no fearsome male oni with a huge iron club. She was shriveled like their own grandmothers.

Fine, they said. Be quick about it.

Weeping, she explained who she was. How, after she killed her lover, her intense jealousy and rage had transformed her into an actual oni. After that, no matter how hard she tried, she could not stop killing and eating humans. No matter how she tried, she was never able to change back.

As she said: "This is the curse of being alive: you get hungry."[4]

The woman spoke of her regret. She asked the audience to copy sutras as an offering for her. And for the men with wives and daughters, she warned: "Do not fail to tell them my story and forbid them to let feelings such as I had to arise in their hearts."[5]

Then she jumped into the fire and died.

I find this ending depressing. The story of the woman from Mino appears in the second volume of a collection of Buddhist setsuwa called *Kankyo no Tomo*, which translates to *A Companion in Solitude*. Setsuwa means "spoken story," and such tales are known for being orally transmitted and purportedly true.

This volume, written in 1222 by a Buddhist priest named Keisei, is unique because almost all the tales feature female protagonists. I've read three versions of the story of the woman from Mino, though none are called "The Story of the Woman from Mino." One is called "How a Deeply Resentful Woman Became a Living Demon,"[6] a title that does not encapsulate the woman's full narrative.

In many of the folktales featuring male oni, they are slain by human warriors. In this story, found in a rare collection of tales focused on women, the oni-baba, after confessing to the humans, dies by her own hand. I wonder what would have happened if she had told her story and *not* leapt into the fire.

Would she have been able to stop doing what she was doing, or would her oni-baba nature have made it impossible to change? What if there had been more than a split second between the confession and the death? What if someone had responded? Mostly I wonder: Are these the only two stories? The one, where you defeat your monster, and the other, where you succumb to it?

———

I am a woman now, the grown-up I was always waiting to be, and yet I still have trouble articulating the duality of my body and my mind, of myself and the monstrous. When my therapist tells me I am in control, I sometimes wonder: But who am *I*?

Thus, I have spent the past two decades as a scholar of my own brain. My hobbies, interests, and occupations return to the psychologist's quip: All research is me-search. To ensure that I am not slipping, I comb my days for signs. Is it normal for me to get this upset? Am I talking too quickly? Would a regular person also have this emotional reaction, or am I blowing things out of proportion? I watch for harbingers of doom.

The thing about hunger is no matter how much you eat, you will always be hungry again. This is the curse of being alive.

Now, in the rare times the Rage appears, it sparks a less visible transformation. First my heart, and then my breath. If I can stop it there, as usual, often no one will notice. If not, a jagged feeling spreads through my body. To the nearest target, usually my husband, I say: *You—you—*

A five-minute barrage of pacing and frustration, my fingers grasping at the air, followed by my apologies and shame. It is not the Rage of the past, more an echo of an echo, a song I recognize.

How can I explain that, at these times, I still feel the pull of violence like a siren call—my hands longing to destroy? I do not have the language. Even now, when many women speak about their therapy openly, when depression and anxiety are part of our conversational tableau, rage is seen as something else. Acceptable rage, for women, is the kind of righteous fury pointed at inequity. The other kind of rage, the kind that harms, the kind that can lash as easily at friends as at enemies, is the domain of men.

Women are told stories about that kind of rage: how to escape it, avoid it, stop blaming yourself for it. It is understood primarily within the context of violence and abuse.

Even in this story, I feel compelled to say that now, at my very worst times, when my hands long to crush, it is plates they dream of, cups and platters crashing to the ground. And do I do this? I do not.

Admitting even this, I fear, might be too much. I am afraid that if I make a mistake in the future—say, forget to pick up my child at school one day—these words will come back to haunt me. I am not afraid I will physically harm my daughter; I know that I will never harm my daughter, or anyone else, in this way. My daughter receives the most patient and tender version of me.

I am afraid of other people seeing the words *bipolar* and *rage* and *mother* in the same story.

This is why, when I speak of rage, I shroud it in metaphor. Metaphor and simile are useful for pretending like you are trying to elucidate when really you are trying to obfuscate. It is like this, I say brightly, drawing a curtain over the whole thing.

———

I've always struggled with the tidy arc of the typical mental illness memoir, the kind whose trajectory leads toward being "better," though the writers usually don't pretend they are fully healed. Perhaps this is because most mental illness memoirs focus on the Very Worst Experiences, so there is nowhere to go but up.

Yet most of my life has been lived outside of either an acute episode or a blissful recovery. My experience of mental illness is a sine wave. It undulates. It never goes away.

After becoming a mother, I talked to my therapist about not knowing how to have the good kind of anger, how to be angry when I should be angry, instead of swallowing it down. What if you just listened to it? she said. Start listening to your anger.

I was skeptical. First, it sounded somewhat—for a lack of a better term woo-woo-y; and second, because to listen to the Rage was to release the Rage.

And what if you did? she asked. What if you intentionally opened the valve, in a manageable way?

I did not have a good answer. The Rage had always escaped and I had to hunt it down, with varying degrees of success; I had never considered letting go on purpose. The idea frightened me. I understood the concept of a controlled burn, but I knew you could not command the wind.

———

"I am [an oni]," writes Japanese poet Baba Akiko. "With a stigma on my body, I sometimes cry out, wanting to sprout."⁷

Beginning in the 1970s, stories of the female oni—such as the oni-baba and its overlapping sister, the yamauba mountain hag—were reinterpreted by Japanese female writers against the backdrop of rising feminism. These figures emerged sometimes as women who "struggle in vain to break free from conventional notions of femininity and motherhood," and other times as "social dissidents who choose to transform themselves into nonhuman beings," explains Mayako Murai, a scholar of Japanese fairy tales.⁸

In the original stories of female oni, the woman's anger is usually blamed for her transformation from human to ogre. Her monstrous form is her rage made manifest, which destroys both others and herself.

We don't know the villagers' response to the confession from the woman from Mino—the story doesn't say. The narrator only tells us that he doesn't know if the village men offered the sutras for her, for the person who told him the story did not mention it.

We know that for so many years, she lived in that decrepit building, filled with shame, and eating boys she'd kidnapped. And when she was finally exposed, when she was about to burn up, she emerged not, as one would expect, to save herself, but merely to tell her story, and to ask for penance, and to warn others of becoming like her. She showed herself, finally, and then returned to the fire she had left.

On my eighth or ninth rereading of this story I realized that though I'd been wondering *how* the transformation had occurred, I had not thought to ask *when*. The moment of transformation began not when she became angry at her lover for abandoning her, or when she shaped her hair into horns, or when she ran away, but when she killed her ex-lover. "After that," she says, "no matter how much I tried, I could not regain my former physical self."⁹

Yet the warning she gives the audience—to the men, for them to pass on to their wives and children—is against emotion, a tale repeated in other tales and plays of oni-baba and other female oni. Beware of the emotion. But the change occurred because of her actions: she'd killed a man. It was not the rage itself that turned her.

———

In high school I encounter, for the first time, a description of the Rage in a book that is not my own journal. At this time, I am fourteen: my life is theater, church, *Lord of the Rings* fandom, instant messenger, and trying to convince my parents to let me transfer out of my public school.

The sole comfort there is my English teacher, who introduces me to *A Tale of Two Cities*, in whose passage on the carmagnole I read, for the first time, something that reflects how I feel:

> *There could not be fewer than five hundred people, and they were dancing like five thousand demons. There was no other music than their own singing. They danced to the popular Revolution song, keeping a ferocious time that was like a gnashing of teeth in unison . . . At first, they were a mere storm of coarse red caps and coarse woollen rags; but . . . some ghastly apparition of a dance-figure gone raving mad arose among them. They advanced, retreated, struck at one another's hands, clutched at one another's heads, spun round alone, caught one another and spun round in pairs . . . No fight could have been half so terrible as this dance.*[10]

In the scene, Dickens is careful to note that the dance included not only men but women as well. Though he is not speaking of mental illness, his vivid physical picture matches my experience of the Rage, of the things that can take over your body: the clutching at heads, the gnashing of teeth, the dance of five thousand demons cloaked in red like the oni-baba.

To see such actions depicted on the page, in someone else's story, filled my fourteen-year-old self with relief, then dismay.

For the people dancing did not receive names. They did not evolve over the course of the book; they merely displayed fury unchecked by logic or empathy. They subjected innocent men to the guillotine. They were dynamic in their movement but static in their evolution. They were out for blood. They *spun round*, senseless, unstoppable, *until many of them dropped*.

ROKUROKUBI

The cursed women whose necks can stretch like tentacles
at night, searching for lamp oil—or blood—to drink.

THE TEMPLE OF THE
HOLY GHOST

When you turn sixteen, your parents build an addition on the second floor of your home. Now, finally, you have the space you've always wanted. Gone the merry-go-round of changing bedrooms every few years so each sister has a turn alone. Gone the ways your parents tried to create a semblance of privacy for the two in the shared room: the fabric curtain your mom sewed around your bed, the loft your father built in the closet. Now you will all have privacy.

The new wing, to the right of the stairs, comprises a living room, where your family watches TV together in the evenings, and beyond that, your parents' new bedroom and en suite. Instead of your parents sleeping a few feet away from you, they live on the other side of the house. They cannot hear what happens at night. You are separated by so many closed doors.

As the oldest daughter, you get to live in your parents' old bedroom until college. You want to paint it black. It's not a good idea, your mother tells you. You always need light. Black paint would make it too dark.

Fine, you admit, she's right. In the afternoons you curl up on the shag rug in the patches of sun. You compromise by painting two of the walls a deep teal and the ones facing the gigantic windows, a light pink.

The light will bounce off these walls, your mother tells you.

Once you enter this room you never leave. You have all the things you need: a sink and your books and the coffee table your grandparents gave you for your birthday, which has two shelves where you hide all your secrets. (Your parents take their children's privacy seriously. You appreciate this about them. If you leave your journal lying around, you know they will never read it.) Most important, this room has the new blue CD player with the automatic replay button your parents gave you for your birthday. You can lie on the floor and stare at your ceiling fan for hours, just listening.

Music is my *life*, you tell anyone within earshot. When you get your license, you drive into the city for shows. You love the way you feel at a concert, the music like fabric on your skin. Your favorite singer is Neko Case, the front woman of the New Pornographers, whose red hair radiates around her face like a halo or a spark. (Two years later, when you try to dye your hair this color, it will come out fire-engine red: not a flame but the thing that puts it out. It will radiate around your face like you belong on *The Bozo Super Sunday Show*.)

When you listen to music your body changes. You also have this feeling when you are dancing onstage in the musical theater group with whom you spend your entire adolescence. You can neither sing nor act but you can triple-time step as your character taps to *Godspell*'s "All for the Best" or kick your petticoated legs up to your ears in Gaston's Tavern. All the parts you play are grace notes, unknown to people who only saw the movie version. It does not matter. Onstage your body provides nothing but uncomplicated pleasure.

What your body can do: split, crotch to floor. Roll, ankles to neck. When you hear the choreographers spit out a sixteen count, your body listens and follows without pause, as if your mind and body are one instrument. Seamless. It is not only dancing, but dancing as someone else, a performative action designed to be performative. Onstage, each step is premeditated and chosen by another. Your body makes no decisions. It cannot be faulted.

So maybe you love the way your body feels as your legs slash through the air, the ache in your Achilles as you spin, or maybe it's that you have to concentrate so hard on the choreography that everything else disappears. You could call this joy, or you could call this dissociation.

(When you are older, dancing at parties, at bars, it will not be the same. Your exuberance will always be interrupted by self-consciousness: looking around to see who is watching.)

In your scenes, the choreographers place you front and center. You have, they say, presence. Your mother has taught you how to let the energy sizzle inside you, how to spike it from your sternum to your fingers. How to follow a gesture with your eyes. When you are older you will think back on the way your body used to be able to move, but at sixteen you do not think about it at all.

Instead, you think about what your body cannot do: make itself taller or smaller or calmer. You have tried all the things. The 2–4–6–8 starvation diet you read about on LiveJournal. The breathing exercises *Teen Vogue* recommends. The girls in your theater group have long white bones and long white limbs. When you form a kick line, you are always the stub at the end. Onstage your energy gives you the spotlight but backstage you are the same. This body, this bag.

You try to contain it, as the world requires. At this time of her life your mother is listening to *Focus on the Family* as she drives the Chevy Venture to and from rehearsals. The hosts talk about the dangers of a daughter's visible bra straps. After that your mother tells you that instead of wearing a single tank top, you must wear two layered on top of each other, so nothing can peek out of your shoulders or at the armpit. Clothing, which is to say modesty, which is to say purity, is the source of many of your fights.

At your new private school, the nurse can at any time make you kneel before her as she measures the inches from your hem to the floor. Backstage at your theater group—also Christian and conservative— you Charleston in front of the costume mom to make sure your chest doesn't bounce too much. Other girls are forced to wear two sports bras. You don't have enough chest to bounce. At church camp, they tell you the exact ways to avoid temptation.

Always the strapping down, the covering up. Think, always, of the boys, and their wandering eyes.

These places all tell you: *What? Know ye not that your body is the temple of the Holy Ghost which is in you, which ye have of God, and ye are not your own?*[1]

You know all about this ghost, the first one you ever met. When you are three your mother comes into your bedroom—the bedroom whose walls your parents sponge-painted with a border of teddy bears and hearts, the bedroom that is all your own because your sisters are not born yet—and sits on your blue bed and asks you if you want to invite Jesus into your heart. Yes, you say. You like Jesus. He is kind. You have no idea about the principle of salvation, but you know that sin is bad and you shouldn't do bad things. You want to be good. And you don't really understand the Holy Spirit. Now, your mother says, it's living in you, too.

There's Jesus in your heart, the Holy Ghost inside your body—the Bible doesn't tell you exactly where—and as you age, the Rage inside your chest, and the sadness in your bones. Your body is so crowded.

So, you get that your body is supposed to be a temple. What you scoff at, when you listen to these women and their flapping concern, is their fear that any boy would want anything to do with you. You should be so lucky. In junior high, the boy you like says *ew* when he hears about your crush. At the lunch table the next day everyone knows. This is how it goes for you every time. You are sixteen and your greatest desire is for someone to see you, by which you mean, of course, for some boy to see you—for isn't that the same thing? But no one does.

(At night, when you count your tiny despairs, you conveniently forget about that boy, and that boy, and that boy, with their tentative gestures and lamb longing. But it is never the ones you want, and so you think of them as nothing. In your journal, you use the same word for them that you use for yourself—*disgusting*—before you forget about them completely. One warped mirror facing another.)

Like the origami birds and boats your ama shows you how to make, you try to fold yourself in half, and then in half again. You want sharp creases, sharp angles, she tells you, placing a square of blue paper in your hands.

Press the fold with your fingernail, she says. You slice across so hard the color wears away and you see the white underneath. But still your edges never line up right. Something is always sticking out.

These two stories are as old as time: the way the adults hide your body because everyone is tempted by it, and the way you hate your body because no one is tempted by it.

At night you find your secret cache of pills in the coffee table and you leave your body. After swallowing the pills whatever rage or despair or desire or hunger is in you loosens from its tendons. You need a break. You are a sixteen-year-old girl and this is the only magic you know.

––––––

In other stories—ones you don't know yet, because you are sixteen and all the stories you know about women are from the Bible or Greek myth—women have been leaving their bodies for a long time. They are called the rokurokubi. The name means "pulley neck,"* for at night, the rokurokubi's neck stretches until it is as long and curling as tentacles. While her body stays motionless on her futon, her head roams out the window, out the house, out the village.

There are two kinds of rokurokubi. The first goes around frightening people for fun and searching for lamp oil to lick.[2] Their mischief is light. Perhaps remaining tethered to the human body, even if only by a mile-long putty neck, reminds them of their humanity.

The second kind, called nukekubi, have necks that detach completely. They look like severed heads floating above the countryside. Fully removed from the body, the nukekubi have no scruples. Perhaps they forget—are allowed to forget—who they are. They go after humans. Bloodthirsty, some stories say. Nukekubi nuzzle their smooth faces into that tender spot between the earlobe and the jugular, and then they bite.

Many yōkai, like the kappa, are born yōkai, but rokurokubi are human first. They are transformed into yōkai as a result of a curse, and cursed as a result of sin. Sometimes it is they themselves who have done something wrong—angering the gods, being unfaithful—but most often the women are punished for the crimes of their husbands or fathers: their men.

For example, in one story, a woman is dying of sickness. Her husband, seeking a cure, is told by a peddler to feed her the liver of a white dog. Their pet dog, incidentally, is white. How lucky! He kills the dog and feeds the liver to his wife, whose health is restored. But when she gives birth to a daughter, the daughter turns out to be a rokurokubi. And when she grows up, she is bitten by a white dog and dies. It was her father who killed the dog, but the dog takes its revenge on her.[3]

---

* Besides "pulley," the kanji character *rokuro* can also mean "potter's wheel."

But your father does not kill a white dog and feed its liver to your mother during her pregnancy. As such, to separate from this body, you must take matters into your own hands. Every night you swallow the pills, which you count out and record as carefully as calories. You measure the dosages and effects in your notebook. Under the pills your body and your mind splinter.

You think yourself Circe, or her niece Medea. Seven pills of red and four of pink. Five of white and three of red. How many yesterday, how many tomorrow—maybe a couple more today. Knowing you can go elsewhere at night makes the events of the day easier to manage.

The important thing to know about both kinds of rokurokubi is that by day they look like normal girls, regular women. Sometimes they do not even know they are rokurokubi—they wake up in the morning, heart pounding, saying, I had the most terrible dream!

Sometimes they know the truth but when the sun is out and they are surrounded by friends and activities, they forget. Sometimes at night they stay asleep, heads heavy on the pillow. Other times when the sun goes down, their bodies slumber but their heads come alive.

In scrolls and prints, the rokurokubi are occasionally depicted with their hair down, like female ghosts. Letting your hair down—the way a proper, fashionable Edo-period woman would not—shows how they have crossed to the other side of propriety, which is to say: of society.

(A year from now, Cori will, at your instruction, hack at your hair until it's only inches long in the back and the strands at the front reach your chin—the classic scene-girl look. You dye it purple, blue, and pink, gelling it every morning so the chunks in the back stand up. Peacock, your little cousin says when she sees you for the first time. In the psych ward when you are required to fill out a cognitive behavioral therapy worksheet with what you like about yourself, you will write *my hair* in hot-pink pen, every other space blank.)

———

You take the pills to shut down your body's shudders but also so you can travel. In a poem you write, you refer to it as a *hypnagogic wonderland*, a world between realms. It is the best thing you can imagine; you can imagine so little. Half an hour after you take them your body turns to mush and your mind starts to wander.

After your neck stretches you visit the family who lives in the silver alarm clock whose light flashes numbers on the ceiling. (This item was purchased by your parents, who thought it might wake you more gently. All through high school you've had relentless insomnia. You see one sleep doctor, then another. It is a physical ailment, easy to ask your parents about. Your sleep doctors prescribe you Lunesta, Sonata, Ambien, Xanax. The pills all go in the drawer. You add them to your mix.)

You hold the clock up to your eye, stare into the hole where the light emanates. Four people live in the clock, and you visit them every night. When the spots of light flicker, you write their messages in your notebook. You cry about the happenings of the clock people, like when the clock daughter has an abortion.

Though you hope this writing will provide some kind of revelation, on the mornings after you only ever find scribbles that undulate across the page like EKG lines. Still, you take the illegibility as a sign that you *were* truly outside yourself. The taking of the pills and the recording of the pills are intrinsically connected. You are trying to make this happen. You are taking the body as far as it can go.

As a child, you learned that when people died and their bodies turned to dust, their souls—which you understood to be the mind, the part of you that thought and felt—ascended to heaven. The body was just an earthbound, temporary space. Later, when you are in college, you will sit in the back row of a philosophy class, drawing comics with the pale boy next to you. Descartes could doubt the reality of his body, but not his mind, your professor tells you. Cogito, ergo sum: I think, therefore

I am. To even try to doubt the mind proves the mind's existence, for to doubt is to think, and to think is to exist, to be real. This makes sense to you. You doubt your body every day.

But now, when you are sixteen, you make your thirteen-year-old sister stay in your room to document what you think of as your nighttime journeys. Don't tell Mom and Dad, you warn her. Cori is afraid you will overdose. She is afraid you will die. She watches over you every night to make sure you find your way back to your body.

———

In another story, a traveling monk is offered shelter by a woodcutter and his family.[4] When the monk wakes in the middle of the night, he discovers the family's headless bodies lying on the futon. He suspects murder until, on closer inspection, he finds no blood or signs of attack. Rokurokubi, he thinks.* If they find him in this state, they will kill him, they will suck his blood.

He also knows the one proven way to defeat a rokurokubi: to move its body. He pushes the woodcutter's body out the window. When the head returns, it shouts: "Since my body has been removed, to rejoin it is not possible! Then I must die . . . Before I die, I will get at that priest! I will tear him! I will devour him!"[5]

The rokurokubi attacks the monk, who defends himself with a tree branch. After the rokurokubi bites onto his sleeve, the monk manages to grab the rokurokubi's topknot and smash it to death.

But death is never the true ending. The monk, for all his strength, could not pry open the jaws of the rokurokubi, and so he continued on his journey with its head hanging off the sleeve of his jacket. Everywhere he went, people ran away from him in fright. He

———

* Generally speaking, rokurokubi are women, but there are a few stories, like this one, that feature male rokurokubi.

could have just removed the jacket, but he began to speak of the dismembered head fondly, calling it an omiyage, a souvenir. He took the head with him wherever he went, until his natural death.

Sometimes it is hard to decipher if what we carry with us is a trophy or a wound.

———

For years you try to rid yourself of the pills. For years you fail. You replace the pills with other kinds of rituals. You try cutting and then burning, but your skin bubbles and the motions hurt and all of it makes you feel more, not less. This is not what you want.

When you are sixteen you think the pills will be the monster you will always be trying to vanquish. Staying clean, you think, will be a lifelong fight. You know nothing.

Only when you are diagnosed and medicated properly—prescribed anticonvulsants, originally designed for epilepsy, that also can stabilize mood—does your body start, finally, to calm, and it becomes easier to live inside it.

By the time you are in your early thirties you will rarely, if ever, think of the pills. That desire will feel like another life entirely. You only take the medications specifically prescribed for you, which you keep high on a shelf out of the way of your toddler and her insistent hands. When you swallow the pills, you never think, one more. When you take a single pink Benadryl for your allergies, the power of it knocks you out. But when you are sixteen and a bottle of what you call *the pinks* lasts only a few days—the idea of one pill laying you to waste is laughable.

When you are sixteen you think, in the future, if you even make it to the future, you will have made peace with this shell you live inside. But your body slowly begins to change over time. You start to feel the muscles in your back, even at rest. You're no longer able to move your legs wider than ninety degrees. Everything softens and these changes make you brittle. Decades in the future, when you lie awake at night, it is this body that you will still think about.

———

As an adult you will love all the music you love now as a teenager. When one of those songs comes on your heart will careen in the exact same way, making it hard to believe any time has passed at all. When you are an adult you will go to a New Pornographers concert at the Aragon Ballroom, thrilled to be attending a show after so long. You drink a lot of water since you get dehydrated easily, and you arrive early so you can be front and center like always. You're so excited that by the time Neko Case comes onstage with that halo of hair, your stomach starts sloshing all the water back and forth. By the third song, you're pushing your way to the bathroom, where you puke up your dinner. You return to your spot and lean against your husband. It doesn't help; you still feel queasy. You go upstairs to the balcony, where you can sit down at a table. Up here people are chatting and sipping beers. You can hear the music, but the band is so far down below, you can barely see them.

As an adult, you will think: Am I the same person as I was? Am I different now? Your mind feels the same. When you feel your arms, they are not the same arms. Your legs are not the same legs. Only your hair is the same, thick and dense and wiry.

By this time, you've spent so long ignoring what your body tells you. In a way, this was adaptive; many of your body's cues were false. Danger danger danger in every situation. The way adrenaline coursed through you at the slightest provocation. You might be feeling this way, you learned in therapy, but it does not reflect the truth.

But what was once adaptive can easily become maladaptive. Those lessons get muddled in your head. You assume your body is telling a lie. Shut down all your desires. Look away. Occasionally, under stress, you forget your body's hunger. You have to set six alarms on your phone: *Eat breakfast (Did you eat breakfast?) Eat lunch (Did you eat lunch?) Eat dinner (Did you eat dinner?)*

Your body will travel down the path it remembers well.

The problem with gliding tenses from the historical present to the imagined future is that it elides the fact that at some point someone— some version of you—has to do the actual doing. The actual making of peace. Which of you will undergo the continual, achingly slow process of learning how to transform, as psychologist Bessel van der Kolk says, the "inner sensory landscape of [our] bodies"?[6]

———

The stories talk about the lengths the families go to rid their daughters of the rokurokubi curse. Some people interpreted the rokurokubi's wandering head as a representation of the woman's detached soul, separated as a result of illness.[7] To be a rokurokubi, or to have one in your family, was often shameful, for it was tangible evidence that a sin had been committed, that the gods were displeased. They were associated with the world of the night: there are stories of prostitutes who turned into rokurokubi while having sex. During the Bunka period (1804–1818, the same time as the Khoekhoe woman Sarah Baartman was exhibited around Europe as the pejoratively named "Hottentot Venus," and a decade before the conjoined twins Chang and Eng Bunker started to perform around America), large crafted figures in the shape of rokurokubi appeared in freak shows.[8] We love to look at a thing from a distance, to say, There but for the grace of God . . .

(But look: In the woodblock prints of the women, in the ukiyo-e, in the scrolls, see the smile of these rokurokubi. Their heads are so far away from their bodies. They are grinning from ear to ear.)

It is easy, in memoir, to say: *I was a rokurokubi and now I am a woman.* To use the reflective voice to measure the distance between the you- then and the you-now with your hands: *that much.*

But you are always sixteen, in that room, on that floor, gripping onto the rug, staring at the ceiling fan looping round and round. You are sixteen and according to neuroscience your brain is still plastic. Not as plastic as a preschooler's, but more elastic than it will be as an adult. Because of this, it's easier to change your mind, it's easier to learn. Your brain still can stretch.

So, listen: If I tell you that the body is not merely those things—not a wound, a trophy, a temple—but a home, a thing to ground into, not separate from, will you believe me?

If I can convince you then, will you convince me now?

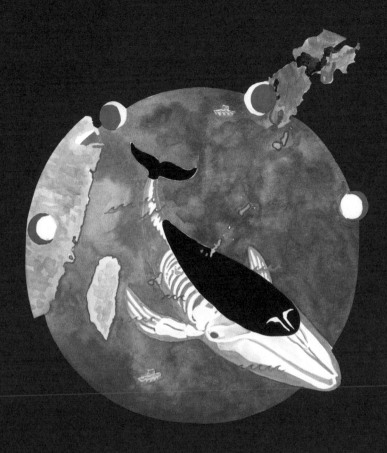

BAKEKUJIRA

The skeletal ghost whale who haunts coastal
villages, seeking revenge for its death.

THE OFFING

It has always puzzled me that most people

seem more intrigued by cetacean death on beaches

than by their life at sea.

—LIAO HUNG-CHI[1]

I

Long ago, in the year of our Lord 2006, in the middle of a warmer-than-average spring in Chicago, time stops. It stops the way a train stops when an animal runs onto the tracks. A screaming brake.

For most people on the planet, outside the zone of influence, this time stoppage has no effect. But those near a certain house in a certain suburb of Chicago, closer to the perimeter, notice strange effects. One woman in a particular cul-de-sac holds her screaming baby for two hours, only to look at the clock and see only a minute has passed. One mother sparrow lays her eggs in a nest in a column of a particular house, only to discover, upon her return, that she has not yet laid them.

In other words, the only ones who notice the shift in time are the people in close proximity to our seventeen-year-old heroine, who at this very moment is refusing to leave her bed.

The audience is invited to gaze upon this lump currently hiding underneath a pink Hello Kitty comforter. Every few minutes, when she flips back the covers for air, you can see her ratty hair, dank T-shirt and gym shorts, her hand clutched so tightly around her journal that her chewed cuticles turn white.

Okay. So at least she's not dead. But then what? Merely *performing*? The sparrow, glancing through the window, cannot tell the difference. She just sees our heroine lying there like a beached whale. *Get up, get up*, she trills. If the time does not accelerate, her eggs will not hatch.

The air around the house shimmers like heat off sand.

Every morning: bed. Every afternoon: bed. Every time her parents peek in: bed. It has been days.

She sweats inside her cocoon, presses *play* on her CD player again. The disarmingly jaunty chords of "The King of Carrot Flowers, Part 1" start once more. Through the window the sparrow has only heard one album, over and over, all these days.

The audience tuts. The question on everyone's lips: What is she doing? Or rather: Why isn't she doing anything?

Every so often her flip phone beeps. *Are you ok?* her friends text. Her boyfriend leaves a voice mail: *Is it the mono again?* She does not respond.

People want it to be mono because people understand mono, which our heroine contracted last fall. You could see the results on a test—the strep test, the mono test, everything positive—and on her skin. The hepatitis of the liver, the penicillin allergy hives that spread like mold on bread. Every square inch of her body rashy, itchy and red.

In the disgust there was, too, a pleasure. She did not have to go to school, and for all her litany of illnesses no one questioned them.

During those mono weeks, her mother held her forehead and let her sleep. Her father knew exactly what to prescribe. Her sister shuffled her homework back and forth from school. Our heroine lay in bed then, as now, but without needing to ask herself: *Are you trying hard enough?* She looked at her test results and knew it was not her fault. No one asked her: *Is this real?* She was not doing anything wrong.

But this spring there is no blood test and its verifiable answer. There is only our heroine lying in bed, and all the people around pushing and pushing and pushing, getting nowhere.

———

The fishing villages of Japan knew what to do with a beached whale: kill it and eat it. Or more often, kill it and sell it, for its flesh was worth more sold for money than eaten as food. There was little else that could be done for these creatures, tens of thousands of pounds, washed up on their shores. Not all villages participated in this practice of passive whaling, but many did.

Though, as with all whaling, you had to be careful. Think of karma. Think of *what goes around comes around*.

Think of the village off the coast of the Shimane Peninsula, whose inhabitants learned their lesson the hard way.[2] One night the young fishermen saw something glowing off their coast, a shimmer in the offing. (Here, *the offing* refers to the far-distant part of the sea, as viewed from shore. Here, it does not refer to the only context the heroine knows, the colloquial term *offing yourself*.)

They wondered what this was, this white shape barely visible through the fog. As it swam closer, they saw it was a baleen whale. They launched their boats into the darkness, excitement fending off the midnight chill, as each man imagined how much wealth he'd gain from selling off the whale's meat and oil.

When they got closer, the fishermen readied their spears, steadied their arms, and threw. These were men born of the sea; they prided themselves on being able to aim true. Yet none of their spears hit flesh. They roped them in and checked the tips. Nothing.

Let's go closer, one fisherman suggested. Others were more hesitant. While they were deciding the whale itself swam nearer, and then they saw what the mist and shadow had occluded: that this was no whale at all. Instead, it was an enormous skeleton in the shape of a whale, its long white rostrum glimmering in the moonlight, reflecting off the dark water. Without its fin flesh, the phalanges skimming through the waves looked like the bones of a human hand. From somewhere inside the skeleton, an eerie light emanated.

Bakekujira, one man whispered. Ghost whale. Skeleton whale.

A humpback whale can weigh seventy thousand pounds alive, but only five thousand pounds when cleaned. Still: five thousand pounds is no easy feat to keep afloat, when you have lost all your fins and all the things that keep you buoyant.

But vengeance can keep one afloat for a long time.

As the fishermen marveled and shuddered, the waters around them filled with fish the likes of which none had ever seen before. Remember: these men knew that sea better than they knew the land, better than they knew their own children. But these fish—these had never been seen in those shores, not by their fathers, or their fathers' fathers.

Look, one man said, pointing up, and all the men tilted from sea to sky, and above them they saw a phalanx of birds, gray blue and translucent.

What is this? What have we done? the men cried out.

Finally, the tide receded, and with it went the shoals and flocks and the bakekujira itself. They rowed their way back to shore, asking each other: What did we do?

———

The longer the heroine stays in bed, the more upset her parents become. She can hear it in her mother's voice, carrying through the door from down the hall. Of course this is happening to her, her mother says to her father. Ever since February, she's been going out almost every day. Of course she's crashing.

And okay: her mother is right about her schedule. Because while neither the heroine nor her mother thinks the word *mania*, in the past semester she's made a whole new group of friends and found her first boyfriend and all her weekends have been spent at concerts. And okay, she's been staying up until three or four in the morning because she cannot sleep, and because she cannot stop instant messaging the boy she loves, a fellow insomniac who unfortunately is not the boy she's dating.

It is so easy for our heroine to fall in love with a disembodied wall of text. It has happened with boys she meets at Lollapalooza and college admissions visits and through LiveJournal comments and the letters they send in response to her zines. Even if she knows them in person, the relationships, such as they are, develop over instant messenger.

What are these boys but archives, and she has always loved an archive. That is what she knows how to love: banter, and a boy without a body.

These boys send her their stories line by line, and when there is a pause, she fills it in. But nothing ever comes of it. She never knows how they feel. The only thing she shares with them are the endless hours, typing and typing.

And isn't that the point, kind of? To dine solely on her own longing? To nurture this feeling that belongs to her alone? To build the shape she wants out of a shadow of a skeleton?

But these things—the boy, the boyfriend, the not sleeping and going out all the time—aren't the sole reasons for time stopping. This whole spring, she has felt she cannot keep up. Her brain, once her staunchest ally, is mushing. In March, she opened a blue book and could not remember anything for her test, though she was the one who created the study guide for the class. Instead, she covered the blue book with lyrics from her favorite songs: "New Slang," "Holland, 1945," "Anthems for a Seventeen Year Old Girl," "Mad World."

Afterward, she wrote in her journal: *I am so tired of being smart Jami . . . I feel free as a bird. I feel uninhibited. I like knowing there is no more pressure. This is probably me being lazy and trying to use some psychoanalytic reason as an excuse.*

She waited for the teacher to fail her, for everyone to notice something was wrong. Except her freedom did not come. Later, her teacher announced her name along with the kids who had been sick. Their makeup tests would be waiting in the library. The teacher did not look at her.

This is what happens when you are very good: people give you second chances. It is easy to ignore what is there.

So her brain was claggy, and her body ever more distant. First, she stopped eating chicken and fish, and then she limited herself to seven foods only: oranges, yogurt, granola, tofu, tea, water, and pills. She placed the water and pills in separate spots on the food pyramid she drew for herself. She was tired of making choices.

Help me, God, she wrote in her journal, all that spring. *Help me, pills,* she said. When you serve two masters your list of desires always dangles on the tip of your tongue, leaving you dumb and open-mouthed like a humpback on a sandbar, uvula flopping *ka-thunk, ka-thunk.*

———

In the village, word quickly spread of the bakekujira. You will not believe this, the young fishermen said.

But the elders did believe it. They had heard of this before. A whale is sacred, they said. A whale can seek vengeance on the village that hunted it. And a bakekujira brings not only a haunting, but a curse.

Long, long ago, there was another whaling village, they said. In that village, there was a whale who transformed into the guise of a human woman and appeared to the chief priest in the dead of night.[3]

Please give me safe passage through your waters tomorrow, she asked him, for tomorrow I give birth.

The other whales in her pod must have told her not to risk the nighttime visit. But she felt the calf swimming in the swells of her belly the way she swam through the Sea of Japan.

It was lean times in that village. The taxes high and the catches poor, the people hungry. Hunger, like poverty and pain, can keep us in an eternal present. The next day the villagers hunted and killed her, though it was a known taboo to hunt a pregnant whale or a calf.

Then bad things began to happen in the village. First the barely noticeable kind, the kind you can attribute to a bad day or bad luck. Then a deadly pandemic arrived, and the village could deny it no more—this was the curse of the mother whale they'd killed.

Upon hearing their elders tell this tale, the fishermen who'd seen the bakekujira paled. Now they knew what was coming for them: sickness and death.

Look: it's not a whale, it's a skeleton.

Look: it's not a whale, it's a ghost.

Look: it's not a whale, it's the future calling, here to avenge the past.

———

The longer our heroine stays in bed, the easier it is to stay in bed. This is how time works.

And here's the thing: Time has not stopped *moving*. It has just stopped moving *forward*. From the outside, both look the same. But underneath the pink blanket, where her sleep is interrupted and filled with ragged dreams, time whirls like a stalled typhoon.

She sees nothing of the past or the future. She sees motion folding back in on itself, each day as bad as the day before as bad as the next. Think: *Groundhog Day*. Think the *was* and *is* and *is to come*. Think of a joke infinitely retold, its humor decreasing each time.

Despair is the sea far past the horizon line. The heroine will never leave this point. Distance, fixed. Aperture, fixed. ————————————————. Just ————————————————. (Some things can't be translated properly, even if you grew up speaking the language.)

It is not sadness it is not grief it is not loneliness it is not being unloved (though it is, too, those things), it is the feeling that no matter what you do, no matter who you are, nothing, absolutely nothing, will change.

I will be dead before I turn twenty-five, she writes in her journal. It is both fear and desire.

It is not death itself that is particularly compelling, it is the idea that she can take a break without everyone asking her why, of not having to feel like this. She knows what people say of mental illness, especially what Christians say of mental illness. She has heard them say it before: *Just pray. God will provide.*

She knows the way people look at the mentally ill: *What did you do to deserve this?*

Our heroine has been taught her whole life that death is not the ending, but the beginning of a life in heaven that is much better than the one here. The thought of death is a warm stone in her pocket. If she ever needs an out, here it is.

———

Look at her mother and father huddled outside her bedroom door.

Should they knock on the door? Should they not knock? It's been days. They worry about interfering with a time stoppage. Won't it disrupt *more*? Couldn't it cause everything to tilt off-balance?

The father still has to go to work, where no one notices the suspension in time. He returns to the house, face drained. The mother homeschools her youngest daughter and ferries her middle daughter back and forth to school, dropping off one instead of two.

Eventually they cannot take it anymore.

In this fairy tale, where each repetition layers like sediment, farther and farther away from groundwater, there was a girl. Let us look at her. She is seventeen.

(*She*: a distancing. *She*: a way to get close. I've tried to tell this story so many times, but when I use *I*—when the girl is *me*—the story sputters in my mouth. Becomes too big or too small or too askew. It is wrong. I promise I am not telling this story like this to keep you away. I want to invite you in, the only ways I know how.

This is what the ghosts know. To get close enough, sometimes you have to pretend you are not you.)

Look at her. Unmoving. Her parents are tired, not of the heroine herself, but of this. The story she is currently stuck in. They sit on her bed. She shrinks. They stand.

We know something is wrong, they say. What is it? they say.

Our heroine has been waiting for this moment for so long—since she was twelve—and now that it has arrived, her tongue is thick and ponderous as a cow's tongue in her mouth.

Her mother is not at a loss for words. She begins to spit them all. Her father cuts her off.

I think you need more help than you've been getting, her father says.

She doesn't take care of herself, her mother says, her eyes uncharacteristically slanted. She looks at the lump under the Hello Kitty blanket that is her daughter.

If you're sick, we should take you to the emergency room, she says. Do you want to go to the emergency room?

Our heroine has never heard this bitter sarcasm in her mother's voice before. The time stoppage has made her mother's jaw unhinge like a nutcracker, scrabbling for something to chew on.

They can't help her in the emergency room, Donna, her father says, his voice now soft. Can I talk to her alone?

Our heroine's mother leaves, and her father sits on the edge of her bed.

Our heroine remembers when she was twelve, when she wrote about mass suicide, and how the school referred her to a therapist. She remembers the therapist, who said, I think she is depressed, and her father, who said, She is my daughter, and I don't think she is depressed. Our heroine did not say anything then. Who was she to speak against her father, who always knew everything?

Our heroine says nothing now.

He says the heroine's name. She looks over her covers, because he says it in his doctor's voice, the one that makes his patients love him and his practice the highest rated in the system. The voice that says he sees you, and whatever ails your body, even if you do not.

Remember that one time on *Everwood*, he begins.

Her family speaks the language of television, often the only thing that can bring them together. It is easier, when navigating the interior hums of familial relationship—no less scrutable to those within than to those without—to gesture toward emotions that can be contained in a twenty-seven-inch CRT. Neither the heroine nor her father is particularly interested in this WB drama about a small-town doctor, but like much of the TV programming in their house, it is absorbed by osmosis.

She is lulled by her father's doctor voice as he tells her the story.

Once upon a time, in Everwood, there were two doctors, he says. One of the doctors had a daughter, and this daughter was depressed. It was hard for the doctor to see the daughter objectively. And it's hard for me. I was viewing you as your father without seeing what you really needed.

Her father touches her hair.

This scene hits all the marks of an after-school special. The moment of the father admitting his reasoning might be flawed. The moment when the heroine, under the covers, starts to cry. The moment when her father, when she emerges, red and snotty, asks, Do you need help?, and the heroine says, finally, yes.

The words they say to each other are not nuanced or philosophical. This is how the story goes. This is when our heroine's trajectory with her parents begins to follow a different angle. Her mother takes a few more days to soften, but when she does, she is there.

Time starts again. The sparrow looks at the heroine through the window, shakes her beak, and says: *Well, it took you long enough.*

———

A little while later, our heroine and her father sit in a waiting room so sparse it looks like it belongs to a shell company. Brown bulky furniture, a ficus. A middle-aged man in a rumpled tweed coat walks in carrying a gallon of milk. After he walks through the inside door to the offices, our heroine's father whispers to her—Whole milk! See the red top?—and they giggle. Five minutes later the milk man calls out our heroine's name.

She goes home with two diagnoses—major depressive disorder and anxiety disorder—and three prescriptions, the first of dozens and dozens. She will record the names of these prescriptions as carefully as all the bands she's seen. She stops her recreational pill usage.

When even this does not do the trick, then it's a withdrawal from school and three weeks in a psychiatric partial hospitalization program for adolescents. She goes back to her high school only once after that, on yearbook signing day.

You don't have to sign today, she tells her boyfriend from across the room, you can sign it later. Though she knows there will be no later. She imagines what they all think of her. It is nothing good.

―――――

The heroine would prefer for this entire story to be excised. Not because she finds the telling traumatic—it has been told so many times that it has ossified—but because none of the versions she's told seem right. And also: once this story is on the page, it will overtake all the other stories. It will seem like the main point. Whereas the heroine thinks that the main point lies elsewhere. But what does she know? She is only the heroine.

But without this story, nothing else makes sense. That's the problem. If time is a long bolt of fabric, then hers has a pin in 2006. No matter how long the fabric, when it is moved it will always undulate around the pin.

II

Spring gives way to Chicago's notorious wet heat. Look at the sun, slipping through the sky. Time, no longer fixed to her body, begins to quicken. It needs to make up for what it lost.

Summer is the season of church camp, theater camp, a family trip to California, a van across the country with her youth group sliding around in the back. Summer is always a good time for our heroine. She looks remarkably well. Rosy in the cheeks. She does not think of death at all, cannot even remember what it feels like to think of it.

That shimmering heat. In the mornings, she's not sure if what she sees is real or a mirage. Sweat drips down the back of her ankles into her gym shoes. Everything slides together.

If it were summer year-round, maybe things would be different for our heroine. But the shadows always lengthen. That's the way of seasons, isn't it? By the equinox it's the pills again. Something about the angle of the sunlight in autumn makes her stomach clench.

This is the point where the audience usually says *stop*. They pull their hair. Have we accidentally rewound the tape? they ask. Hasn't our heroine learned? Can't you compress this timeline so we don't have to watch the same interminable loop?

The timeline has already been compressed so much, but the audience loves simple arithmetic. They want to know *what happened* and *what led there*, as if there is a single thing that happened, and a single thing that led there.

If they're looking for concatenation as clear as the clack of an abacus, they're watching the wrong kind of show.

———

Long ago, in the year of our Lord 2006, in the middle of autumn in Chicago, time blips. Our heroine is now a senior, though she isn't attending an actual high school. She's taking classes online and at the local community college.

So many medications, so much money for her therapy and her partial hospitalization program, and here she is, back on her bullshit. Every day more pills (look at them, accumulating like beetles on a buried log), every day less interest in what could happen afterward.

Our heroine has spent so long flirting with the edge that she can't see when she's toppling over it.

One night she returns home from a college visit and gets into an argument with her parents about her virtual AP U.S. History class. She wants to withdraw from the class; her parents do not want her to withdraw from the class. That is the straw.

Inside her bedroom every part of her is activated. (*Once the Rage is in her body, she cannot easily get it out.*) She counts out twenty-five pills— she likes a nice round number—and writes down the total milligrams in her journal. Hungry as an oni-baba, she swallows them all.

What is the heroine trying to do? Members of the audience— including the heroine, when she rewatches the tape—have struggled to parse this out. The heroine wants to die *generally*, but at this moment is she trying to die *specifically*? Is it active or passive whaling? On this point, the archive contradicts itself.

Our heroine is seventeen. What is grief to her, what is loss? Her father is not dead yet. She thinks if she does not choreograph death herself, it will not happen.

Our heroine's fingers begin to feel tingly, then her feet. She lies down.

What finally scares her is when she cannot feel her limbs. The other symptoms—the headache, the blurry vision, the falling down when she stands—she has experienced before. But now it feels like parts of her body are unplugging from their sockets.

She instant messages a friend that she took a lot of pills and does not feel good, and her friend says she's so sorry, and also that her mother is calling, and she has to go. *bye*, the friend says.

She calls the suicide hotline but hangs up before anyone answers.

What to do, what to do.

October 27, 2006, she journals, in huge, jagged letters.

I can't feel

my body is numb

I'm sorry

Lord make this go away make me better please

She calls her sister, who is at a church event forty minutes away, instead of her parents, who are watching *Numb3rs* down the hall. Cori can tell something is wrong.

I took a lot of pills, the heroine says. More than usual.

Do you want me to tell Mom and Dad? Cori asks. Even now, Cori is seeking permission. This is how the heroine has trained her, this younger sister, now barely fourteen, who has been keeping watch as carefully as she has been keeping secrets.

Our heroine thinks as best she can through the fog. On the one hand: her parents think she is better. She can imagine her mother's face crumpling like old newspaper. On the other: her body feels like it is decomposing.

Do you want me to call them? Cori repeats.

She shuts her eyes. Zoom in on her face, as two things slowly dawn on our heroine—

first:

that she might actually be dying,

and second, the more crushing of the blows:

that all her recent actions point to the fact that she does not actually want to.

See the realization spread from her eyes to her mouth. See the horror.

Call them, she says, as she climbs into bed and waits for the sound of the call that will tell her the end of the world is nigh.

————

Her father carries her into the van like a fireman, with her mother and Kristi scrambling in after. What's happening? Kristi repeats from the last row. She has recently turned eleven. She is in her pajamas.

Our heroine's head is in her mother's lap, her mother stroking her hair.

Were you trying to kill yourself? her father asks, driving through the night.

No!! the heroine says, exclamation point exclamation point, and even if it wasn't true when she took the pills, it's true now. She grieves it, this door now closed to her.

If not this, then what?

Existing?

Forever?

————

Today we still struggle with what to do with beached whales. In some areas of Japan, communities mourn the whales that wash up on their shores. In other countries, people try to dispose of them as quickly as possible, and this comes with its own problems.

In Taiwan, the body of a beached, decomposing whale corpse is transported from the shore to a research center.[4] Over six hundred locals come to watch the leviathan body move through the city's street vendors and betel nut shops. We love to watch a secret thing exposed.

The truck is rumbling along as usual when the whale explodes. Blood and blubber sprays the onlookers; chunks of flesh shoot over eight hundred feet away. Gas has built up inside the carcass.

Fearing a similar situation, when a beached whale dies in the Faroe Islands, local authorities call a marine biologist to release its gasses safely.[5] What they want: the slow deflation of a balloon. But when the biologist spears the underside of the whale's belly, he accidentally hits a gas pocket. Entrails explode out of a gash that looks like a gaping mouth. This one is marked for posterity in a fourteen-second YouTube video entitled "Sperm Whale Explodes."[6]

And in March 2011, fifty melon-headed whales beach themselves on a shore in Ibaraki Prefecture on the north coast of Japan. People are puzzled. The whales are perfectly healthy. Six days later, no one remembers the whales, for just one hundred miles away comes the tsunami that kills almost sixteen thousand people.[7]

A few scientists claim that whales can detect geomagnetic anomalies,[8] though the research is inconclusive. But it's something: whales as harbinger of doom. Japan is the most tsunami-prone place on earth, receiving, on average, one every 6.7 years.

———

As an adult, the heroine learns that her family remembers all the details of the night of her overdose as if it happened yesterday. In the versions authored by her family—in which the heroine is not the heroine, nor the narrator—this is the night they return to. Her mother remembers exactly what they were watching on TV. Her mother thought her daughter would die. Her sister Kristi wrote about it for a school assignment. Her sister Cori remembers it as the night that shaped all their dynamics afterward. They remember this night better than she does, the way trauma has engraved it.

She is only the heroine of one story. But all the members of her family have to live with what she does for the rest of their time on earth.

What haunts her is the way all their paths afterward forked. She can think about the event itself but the rifts it created, decades later, get muddled in her mouth. How to tell that kind of story?

If she could see the beginning and the end, she thinks, she could figure out what it means. She wants the story to be done with so she can stop thinking about it. (Look: it's the future calling, come to avenge the past.)

But the story is not finished unless we stop changing, stop transforming, which is to say: unless we are dead.

———

Unlike many other yōkai tales, which proliferate in many places under many names—the kappa, the kitsune, stories which you could tell until you die of old age and still not finish them all—there appears to be only the one story of the bakekujira, though there are others about ghost whales without the skeletal appearance. It is a yōkai known more in contemporary times because Mizuki Shigeru, creator of the popular manga *GeGeGe no Kitarō*, immortalized this yōkai in his own work. Afterward, versions proliferate based on his version. My translator and I even have trouble tracing it back before him. Did it spring fully formed from Mizuki's head, like Athena out of Zeus?

But Mizuki said that in the 1950s, when he was working on a different story about the bakekujira—living near the coast and consuming a lot of whale meat—a fever overtook him and he became very sick. It was, he said, the bakekujira's curse. The illness only left when he quit writing the story. The whale's revenge.[9]

When you see the bakekujira you know illness is coming. You spend your life looking for the ripples in the water. Is it a whale or a skeleton? Will you be fed or dead?

An event in the past does not stay in the past. This is what the people who have encountered the bakekujira know. The past can come back and get you.

The whales returned to haunt not only the men who killed it, but the entire village, all of them intertwined in their own ecosystem.

After being decimated by pandemic, the villages haunted by bakekujira and other ghost whales implemented what many others across Japan already knew: that the way to appease the whales was to honor them. The villagers haunted by whales built a kuyō tower in honor of the whale's memory, where its spirit could repose. Each time, this appeased the whales, and the illness ceased.

Each village haunted by a ghost whale had to learn this themselves.

The practice of mourning whales was widespread across Japan during the era of classical whaling.[10] Villagers would bury the whales' bones or place them standing up, create gravestones, and conduct funeral rites, praying for the souls of the whales to rest in peace.

In some villages, they listed in their shrines the names of every whale they hunted, alongside the names of their own ancestors. These rites happened in active whaling communities, also when fishing communities simply captured a beached whale. Even though they were not actively *trying* to hunt the whale, they saw the result as the same: a dead whale, a whale to be mourned. The mourning rituals functioned partially as atonement to prevent future retribution, but the hunters also felt real sorrow.

Even after being haunted, the villages did not think, We must stop hunting whales, we must prevent those deaths. This was their livelihood. Instead, they acted, and then they mourned those actions. They wished for the whales to reach enlightenment.

Sometimes we do what we do to survive.

The whale is dead. The bones remain.

KITSUNE

These fox yōkai can be wild shape-shifting
trickster or intelligent divine gods. Or both.

POSSESSION

—

I

Daughter—

One day you're going to ask me what it is like, and I'm going to have to figure out how to tell you without frightening you. I want to show you that I am someone you can rest your head on while knowing there is a chance that this might not always be true.

When other people ask what it is like, they mean capital *B* Bipolar. They want to know about the peaks and valleys, the hospitalizations, the careening. What they want, these other people, is to hear a good story. They want to know how I got better.

But when you ask me, because you are my daughter, you will mean *How will you* stay *better?* And: *Will this happen to me?* And also: *If this happens to me, how will I endure it? Will I endure it?*

And how can I explain this to a child?

———

I will begin with kitsune because you already know about kitsune. We are on solid footing here.

In the Japanese zoo book my mother gave you for your first birthday, the kitsune is surrounded by the other animals: kirin, zō, saru, panda. This kitsune has sweet little eyes and a shit-eating grin. The kitsune, this book tells you, is of this world.

Kitsune is the name of the fox animal and the fox yōkai both, and sometimes we cannot tell if we are looking at one or the other. But *all foxes have supernatural powers.*[1] In that case, does it matter which is which?

The kitsune can hoodwink you out of your meal. They can present as beautiful women. They can be dangerous or not. They can be devoted wives of human men and devoted mothers of human children. They give abundantly to those who show them gratitude. They can be vengeful or playful. Some are funny, some impudent, and some so wise they grow a second tail when they turn one hundred and another for every century after that, until their tails fan out from their rears like a peacock's feathers on display.

And they are also responsible for kitsune-tsuki, which is usually translated as "fox possession."

When a person is possessed by a fox, they begin to behave in violent and illogical ways. They begin to exhibit symptoms of mania or psychosis.[2]

It's kitsune-tsuki, their family members say, recognizing the signs. The name places the symptoms within an explicable lineage and framework. Then they panic, for families possessed by foxes could be ostracized by their communities. To grasp is different than to empathize.

"When a kitsune possesses an individual, it is often in retaliation for something done to the kitsune—killing one of its family members, for example," writes Matthew Meyer.[3] To be possessed by a fox is the victim's fault.

———

In the days of the foxes, the possessed were kept in their homes, out of sight. In my day, we were brought elsewhere.

This is how they thought we would get better, by taking us away from everyone who knew us.

I say *they*, though I chose this path for myself. The morning after I overdosed at seventeen—the morning after the charcoal I drank stopped the pills from absorbing too far into my body—I told my parents from my hospital bed that I did not want to go home. If I went home, I knew I would do the same thing again. Nothing had changed; I had not changed.

The doctor admitted me into the psychiatric ward located a floor above the partial hospitalization program I'd attended the previous spring. While that had felt like a slightly uncanny day camp, the inpatient ward—which I later learned was referred to as *Upstairs*, after its literal location—immediately felt different. Upstairs had no cell phones and few windows, locked doors and bone-white walls.

When I arrived the social workers looked through my duffel bag, confiscating my bras and my pens and my spaghetti-strap tank tops and leaving my Bible and photos and pajamas.

After being stripped of these possessions I hugged my parents in the hall. My father kissed my forehead. My mother and I cried, me noisily and her silently. We'll be back soon, they said.

I cried again that night after lights out, full-bodied shudders full of phlegm and verve. Feeling sorry for me, or for my roommate, the social worker on rounds let me sit next to her in the hallway. Once I managed to remove my face from the wad of tissue, I told her: I chose to come here, but I think I was wrong.

It gets easier, she told me.

I did not believe her. This was what the adults always said.

———

The next morning I tried to get my bearings. From a certain angle—if you squeezed your eyes mostly shut—you could mistake my bedroom for a dorm on move-in day: two beds, two dressers, an overhead lamp, thin carpet.

My roommate, Marissa,* a freckled heavy-lidded thirteen-year-old, showed me the ropes, or lack thereof. In our room the blinds had no cords, the windows wouldn't open, and every piece of furniture was bolted to the ground.

And the door doesn't latch, she said, opening our en suite bathroom door. Inside, a tiny shower was jammed next to a toilet and sink. Don't worry, they don't watch us while we shower, she assured me. Unless you're shaving. Then they sit outside.

I don't think I'll shave, I said.

Their razors are really crappy anyway, she said, letting the door close. It swung back and forth like a pendulum, in tinier and tinier arcs until it stopped. Those pink Lady Bics.

The entire architecture of Upstairs was designed around the ability to be watched; our bedroom doors lined four walls facing inward at a central nurses' station, their own panopticon. The doors were kept open except at night, and even then, the long windows meant they could see us at any time.

The light from the hallway comes in all night, Marissa said. It's really hard to sleep. But you can ask for a pill.

---

* Names and some identifying details in this chapter have been changed to protect privacy.

This was the light I would learn to journal by—each entry in purple crayon, always with an eye to the window, watching for the shadow that meant getting caught.

Upstairs did not have the antiseptic smell of the emergency room or the overcooked-food smell of the nursing home where I used to work after school. Instead, it smelled of absolutely nothing at all. White walls, the absence of scent, and no sound except talking. At home I was never without music but here: blank.

For you I am trying to do the thing that is hard for me: to think about what a place looked like. "Any story relies on negative space, and tradition relies on the negative space of history," writes Matthew Salesses. "The ability for a reader to fill in white space relies on that reader having seen what could be there."[4]

When I try to fill in this white space, to explain what Upstairs looked like, perhaps I am hoping that you will be the kind of person who does not understand this tradition. That you will be the kind of person who does not understand what could be there.

————

Upstairs, *my* Upstairs, was a girls' world. Except for the hour-long morning and evening process groups, we kept away from the boys. We ate our meals in separate rooms. Our bedrooms were in opposite halls. The staff's fear of us having sex was second only to their primary worry that we would off ourselves.

What did *we* look like, the girls who, for a time, lived Upstairs? Like we were twelve. Our appearances limited by what was allowed, and nothing was allowed. We wore old gym shirts and sweatpants with the strings removed. We didn't wear makeup. Our faces were oily in the T-zones and dry in the cheeks and our hair was unkempt. We did not look like the girls in *Girl, Interrupted*, then my only cultural touchstone for a psych ward.

But the actresses in *Girl, Interrupted* were women in their early twenties. We were actually girls.

Girls like Kit Vicious, who I met on my first full day when a voice that sounded like it had smoked since birth carried through the hallway. (Later, I'd learn its scratchiness was because she was always shouting, usually with delight.) After the voice appeared a short girl wearing a hospital gown on top of a Hot Topic T-shirt and a pair of zombie pajama pants. Her hair, chopped off at the chin, was dishwater gray from the purple Manic Panic dye that had leached out.

New girl, she barked at me during my first process group. *Why are you here?*

I don't know, I said, though I did know, at least a little. What I did not know was the lay of the land, the posture I was supposed to adopt to fit in. Apathy, I thought, was the best choice. I'm just here, I said.

Well, slap me like a summer sausage, said Kit, smacking the side of her chair as she spoke. *Aren't we all.*

Girls like Kit and my roommate, Marissa, were the kinds of girls I would later see skanking at ska concerts in VFW basements, or at the Spencer's in the mall, or loitering at the California Pizza Kitchen. But on the other side of an invisible boundary were girls like the seventeen-year-old Tory, who was friendly in a startling sort of way, and who I would only ever encounter in spaces dedicated to mental illness. Her eyes zipped around, landing on one person and then the next in quick succession, and other times staring at nothing. Her words spewed as if from a Super Soaker with its trigger held down.

I knew two things: that I liked her and that I would only like her within these walls. She always interrupted. She couldn't stop talking about Conan O'Brien (*I would have sex with him, and I don't even like sex!*) and refused to pretend to be tough (*What do you do if your parents do something that pokes a hole in your heart?*). She screamed at a girl over a bag of pretzels, and had threatened to stab her mother, but the real reason was the sum of all those things: that Tory would, wherever she was, be clocked as weird. Off. Abnormal.

The boundary between the two kinds of people you meet in a psychiatric facility is sometimes ascribed to the difference between high-functioning and low-functioning. When you are what people term *high-functioning,* you can usually pretend to be in the realm of the well, as I could. People are often surprised when you reveal your mental illness. Wow, they say. Good for you. When you are low-functioning, there is no act of revealing: it is revealed.

The designation is used as a static identifier, as if it is inherent to us or our illness itself, rather than our circumstances and levels of financial and familial support. The terms have no clinical or psychological basis; they are, instead, a social marker. Such differentiation is a trap. They are often used as shorthand for the good mentally ill versus the bad mentally ill.

As always, I wanted to be good.

(When I look up all the girls, fifteen years later, Tory's is the only name that returns any information. Her LinkedIn informs me she is now a social worker.)

———

Emotions ran high Upstairs. We pissed each other off, especially during process group, when we'd get mad at each other for the ways we were managing our own lives. At every meal, though, we would come back together. Marissa was bipolar, and Kit Vicious was bipolar, so they named our group of girls—the ones who ate lunch together—the Bipolar Bears. They didn't mind that this wasn't my diagnosis, they only cared about the way the puncture marks in our stories lined up, like assembling a sheaf of three-hole punched paper.

I suspected then that I was bipolar, though my psychiatrist still thought I had anxiety and depression. After I was discharged from the hospital this would become a recurring argument with him, as I took dozens and dozens of medications that did not calm my body or my mind. He kept telling me to wait. Bipolar is a very serious disease, he chastened. He did not want to diagnose me until I turned eighteen or we had more proof.

His response was similar to my father's when my therapist told him she thought I was depressed in seventh grade. They did not believe what they could not see, these men of science. It took years for my father to recognize it for himself.

When my psychiatrist continued to stall, my parents told me I could get a second opinion. The next psychiatrist diagnosed me with bipolar I right away. He prescribed a medication that was initially formulated for seizure disorders but was often given off-label as a mood stabilizer. It soothed the Rage the way the antidepressants could not. It unclenched my fists.

———

The old joke goes: *When I got diagnosed with bipolar, I didn't know whether to laugh or cry.*

Upstairs I learned you could laugh at the things that also curdled you. Once I tried to slit my wrists, Kit told us at dinner, and then I was like, shit, this hurts.

The laughter that came from communal understanding, even stemming from pain: my father intrinsically knew the value of this, and so did my sisters; for me it was a skill that when acquired still felt as artificial as a magic trick.

Each of our laughs had a different song. Kit's was the *hyuk hyuk hyuk* of Jughead Jones. Marissa's happened all in the nose, a bemused snort. Tory's was alternately thin and reticent or explosive, laughter so close to screaming. Each song included a different minor note, like the splash of beaver musk perfumers add to their solutions, that hint of disgust.

When Kit Vicious named us the Bipolar Bears, I laughed, and when we ate dinner together, we laughed, and this laughter was not despite our diagnosis but directly because of it. I felt free to laugh because I did not need to perform my sadness. I did not feel the need to prove I was sick, like I did with people outside who thought my illness was an attention grab or an evasion of responsibility.

I say, *Upstairs was where I learned*, like I was there forever, though it was only a week. The experience, so unlike the regular tune of my days, made an outsize impact due to its discordance.

But it also felt so long because time in the psych unit was unstable. You had very little control over the length of your stay, and the kids who were involuntarily committed had no control at all. When you do not know when you will leave you do not know how to mark your days. You do not know how much energy to expend on making friendships. And yet friendships bloomed anyway. We were stuck. We had no phones, no Internet. Besides our one nightly movie—chosen by group consensus from a box of videotapes, the cream of the crop of which was *Grumpy Old Men*—there was nothing to do here but talk, and nothing to tell but our stories and our jokes, and listen for the tremor that ran beneath them.

The relief from not having to pretend you were not unhinged, that you were not falling apart. You could just fall apart. Upstairs, the girls told each other why we should live. If we could not believe it for ourselves, we believed it for each other.

———

I cannot remember what the social workers told me or what our cognitive behavioral therapy worksheets said. I do remember Kit Vicious waking me up and telling me that tonight was dirty limbo, her gathering all the girls to play the game, all of us singing *Every limbo boy and girl . . . all around the limbo world . . .* before the night aide found us and threatened Kit with the booty juice.

I remember our collective disappointment at not being able to have caramel apples on Halloween—the sticks deemed a danger—and how we banded together to demand we watch *The Nightmare Before Christmas*. Kit Vicious, who every day in expressive therapy wrote a new zombie story, each more engaging than the last, led this charge. It's not *bad*, we argued. It's a children's movie, we argued, heartwarming, wholesome. In the end it wasn't a matter of morality but of logistics. They couldn't get a copy of the VHS in time.

I remember these things not because they are beyond the pale but because they are the same kind of things any bunch of girls might do at a sleepover.

What I want to tell you is that if this is your name there is joy in it, too.

The night before Kit Vicious was discharged, we gathered in her room and did the Hokey Pokey in her honor. All the girls crammed into Room 6 and sang the words as a group. Kit made up her own lyrics. *You put your hospital bracelet in, you put your hospital bracelet out, you put your hospital bracelet in, and you shake it all about . . .*

The next day Kit packed her things in her SpongeBob SquarePants backpack and left with her new foster mother. We told her not to screw up. We watched out the window as she walked to the van.

Two days later I, too, was discharged. They warned us not to contact each other afterward. It's not good for your recovery, they said. You need to focus on *you*.

We were brought together only when we were locked away, in those rooms where trauma could be meted out as easily as it could be mitigated. Afterward, they thought we should have a supportive community that did not include other people like us. They wanted us to be surrounded by healthy people. Except most healthy people did not know how to talk about illness, or how to talk to us at all. I heard you went crazy, a friend said a month after I was discharged. Everyone said you went off the deep end, a former classmate typed over instant messenger.

At home, I hung up the collage Kit had given me, a black crow against purple construction paper. *The Crow* was her favorite movie, and she used the bird as an emblem on all her drawings. Her credit's in the corner: *Kit Vicious, 2006.*

I left still seeing the same psychiatrist, the same case manager, but those girls I never saw again.

II

I knew I was lucky. Marissa, upon discharge, was sent by her parents to a months-long wilderness boot camp to manage her anger problem. Marissa, the softest-spoken girl, the gentlest girl. Kit Vicious went home to a new foster family. I had parents who came and cared and took me home at the end. After I was discharged, I went to the partial hospitalization program for another four weeks, and when our insurance stopped paying, my parents covered the rest of the days out of pocket.

The staff Upstairs liked to emphasize individual choices, how we could change our thoughts and behaviors, but not the fact that each individual good choice can only do so much in a system designed to forget about you. Like I said: money meant you were remembered.

And at home I had my family, all my other girls. I was used to running in packs of girls, as the oldest of three sisters born to a woman who was the middle of three sisters born to a woman who was the youngest of four sisters.

These generations of women were inextricably linked. Sometimes I even forgot where we began and ended. For example: every time we heard a joke. It wasn't the tone or pitch of our laughter that was the same, but the way our laughter ebbed and flowed, the way we threw our heads back and clapped our hands, wiped our eyes with our pointer fingers as we came down to neutral, and sighed at the end, like we'd finished a feast. This was how you laughed like an Ige woman.

But in those days, after I came home from the hospital, I was vexed at the way the generations above me reacted. My grandmothers had their own ideas of how to help. Ginger, said one, pressing a light-green bag of sugared ginger from the Chinese market in my hands. Pressure points, said the other, gripping my wrist. Do you want peppermint oil, do you want this sour cherry juice? I stopped mentioning in front of them if I wasn't feeling well, even if it was only a headache. I did not see why they could not leave it alone.

I did not have you then. I did not know then the worst kind of love is the helpless kind, how you will throw anything at a wall to see if it sticks, just to feel the swing of your arm.

In lieu of adults, I had my sisters. This was lucky, though the luck depends on whose perspective you're possessing. Cori, age fourteen and fifteen: sleeping every night with her sister, making sure her sister did not die—this could not be called luck. Kristi, age eleven and twelve: on the cusp of becoming a teenager, with all attention in the family diverted to me and my crises. Was that luck?

———

For so long I wondered if I possess the bipolar, or if the bipolar possesses me. Who controls the other, and who takes responsibility? How much can one transform?

Then I read H. Yumi Kim, a historian specializing in early twentieth-century Japan, who says the term *fox possession* is not the most accurate translation for kitsune-tsuki. A better term would be *fox attachment*, she explains: "'Tsuki' conjures a form of superficial contact, one that occurs at the surface, as if fox spirits and humans are two surfaces that stick to one another, rather than a deep and penetrating form of ownership, as the English word 'possession' suggests."[5] She uses attachment as a framework to understand how these episodes relate to family and kinship ties in the Meiji era, and the relationship between these ties and the state. In her work, she studies these fox attachment narratives "on their own terms."

Two surfaces that stick to one another. Not one sublimating the other, but side by side.

———

Stories of bipolar often center on our worst moments, crisis's hot center, even though hospitalization is only one small facet, the tip of the iceberg. When people ask what it is like I never think of my worst moments first.

What it is like: To doubt yourself. To always doubt. To doubt the beginning of it and the end. Is this me, or is this my illness? Am I sick, or am I lazy? Am I happy, or am I manic? Once, I received so much good news in a month that I was dizzy with delight. Is this normal? I asked my therapist. Would a normal person be this excited?

The hospitalizations themselves do not haunt me. The tip of the iceberg looks like nothing. It's the underbelly, the deep traumas, whose impossibly slow passage and melt is felt through its reverberations, throughout generations. The choices I made and make, and how those ripple out in circles and change the landscape.

———

The first time I tried to tell this story I took out all the parts about our family because I didn't know how to make them fit. And also: looking at them hurt. It's easy to tell a story about girls who only existed in my life for a week, for one frame, instead of the people to whom I am tethered for life.

Foxes are solitary, but kitsune-tsuki, fox attachment, affects the whole family. The taint of it taints the whole family. These days instead of *taint* I would say *trauma*, though I am still unsure what is trauma and what is the disorder itself. In some cases, I cannot parse the threads, and in others I resist the word because it feels like essentializing, like saying the disorder itself equals trauma.

This, of course, is reductive. Bringing these things up again feels profoundly destabilizing. I feel a reeling block. I am supposed to be writing this book about carrying your ghosts but some of the carrying is painful. Things you thought had healed crack open again.

What was I really thinking then? What were my parents thinking? I only know that at one point, in the psychiatric ward, my psychiatrist said, exasperated: *You always cry, and your parents always give in.*

This was true, and yet his point brought up only rage, for I thought of the previous seven years and all our family's arguments and how, I felt, I had to fight for everything, and how much of it was unsuccessful. I was angry that he was framing my tears as manipulative when I could not control them.

After the hospitalizations, after the diagnosis, my parents' tight reins loosened. They became more liberal, more understanding, and their relationship to me changed rapidly: from nemesis to bulwark of support. However, a ship that turns quickly leaves much in its wake, and my sisters were left to pick up the pieces. Though I'm only beginning to see the edges, this was the time when our family dynamic altered. My father was still undeniably the captain of our ship, but now my emotions were the weather that had to be avoided.

I do not know enough of this story, or of its characters, to tell it yet. For this, there is the future.

———

Once, when talking about the likelihood of you developing bipolar, which is about 15 percent, my husband, your father, said, Well, we'll just have to hope she doesn't get it.

This is what I, too, hope for—I wonder if I will pass it down to you like the heavy Nakamura eyebrows, the Ige marker for Alzheimer's, the Lin energy—yet I grew irritated with him for saying it.

He didn't understand. Don't you wish you weren't bipolar? he asked. No, I said, and as the word popped out of my mouth, I was surprised to find it true.

By now it's cleaved to me. In each episode—every time I feel depressed, or out of control—I wish it were not happening, that I was not like this, but it shapes my thinking, my mind, my life, and I cannot imagine it otherwise. What I want is to be bipolar without having any episodes ever. Bipolar that is perfectly managed. I know this is not realistic.

Still, it is one thing for it to be mine and a whole other thing for it to be yours.

————

When I went to Japan, a decade after I left Upstairs, I traveled to Fushimi Inari shrine in Kyoto. While kitsune live all over the country they gather in certain places, like this shrine, or in Izumo Province.

Most Shinto shrines are guarded by komainu, paired statues of lion dogs, but the shrines to Inari—the Shinto kami of fertility and agriculture and rice and wealth and sake and tea and prosperity— are guarded by kitsune, for they are Inari's chosen messengers. The knowledge of the divine kitsune develops with age. They can grow a second tail when they turn one hundred years old, and then another every century, until they've accumulated nine. These kyūbi no kitsune are the wisest of all.[6]

Sometimes, the god Inari is described as also being a kitsune. (Here in Chicago our only communion with Inari is when we make and eat the inarizushi named after the figure. When I show you how to stuff the sweet brown tofu wrapper, wet from the can, with the rice we mix with the vinegar powder in the bamboo tray, the same way my mothers and grandmothers did. We call them both *inarizushi* and *footballs*, for their color and shape.)

At the shrine I stared at the rows and rows of red torii of all sizes diminishing in the distance, far beyond what the eye could see. At the shrine, you could buy small flat pieces of wood cut in the shape of a fox face. Visitors wrote their wishes and prayers on the front and left them at the shrine for Inari to answer.

I did not buy a fox face. I was not yet at the point of throwing anything at the wall to see what would stick.

———

Daughter, when you ask what it was like, you know you are asking what it *is* like, don't you? When I press your head against mine. When our bodies are next to each other, warm. Each day a cascade of mistakes and promises to do better—How do we know what bits will accrete into trauma and what will be forgotten in the detritus of childhood? The terrifying part: that as you age, the cycle will begin anew, only this time it will be me on the other side. How to halt it in its tracks? What is the work that must be done: Is *this* the work? Is *this*?

———

The term *Hyakki Yagyō*, the Night Parade of One Hundred Demons, can refer to either the Japanese historical parade of yōkai or as a kind of yōkai collective noun, like a *murder of crows* or a *misfit of ghosts*.[7] An English translation is often "pandemonium" or "riot."

When there was such a yōkai riot, the villagers knew to be afraid.

When authorities use the word *riot*, they are also telling their audience to be afraid.

Growing up my aunts called each other a riot in the same way they'd call each other a hoot. The way when, together, they couldn't stop laughing.

When my sister Kristi—your aunty TT—was a little older than you, she told the same joke at dinner every night. This was when she still sat in a high chair at the head of the table. I remember her telling it with a baby-size fork in one hand, a spoon in the other. The joke, which was only ever directed at my father, went like this:

Knock knock.

My father: *Who's there?*

You.
You who?
You can't say yoo hoo, you're not a girl!

After delivering the punch line, Kristi would throw her head back and squeal with delight. She knew that my father, by his very existence, was slightly absurd: the one male in a household of girls. She told this knock-knock joke to him every single night for months. When I, age ten, asked her when she would stop telling the joke, she looked at me sternly and said: Never.

But then one day she stopped telling it.

Which is to say we grow out of things. The things that were funny once aren't funny later, and, more promising: things that weren't funny once can sometimes grow to be.

From the time you were a baby you would laugh when I laughed, but it wasn't until you were two that you got distressed when I cried. Before that you would look at me quizzically, and continue on with whatever you were doing.

When you turned two, we were sitting on the bed together and I began to cry. I cannot remember the reason—it could have been anything. Me help Mama, you said, and you came over to rub my knee. And I worried that you would grow to feel that you needed to take care of me instead of the other way around.

Right now, you are a child, a toddler, and when your anger burns, I say out loud, She is three. Inside me, though, sometimes I wonder. I don't know how to tell normal childhood frustration from the signs.

During the pandemic I took you to Norridge Park on a playdate to try to acclimate you to other humans. The circle of your world, besides your father and me, was my mother and sisters. You loved adults and feared children. At the park you did not want to play with the other little girl. You ran away toward a slide. As I was chasing you, you stopped and pointed and said, Oh! Underneath the slide, you saw a rust-colored fox with its tail curled around its body, drowsing in the shade.

Stay back, I warned you. Kitsune love to enchant humans.

But you were much less afraid of the sleeping fox than you were of the little girl. Inu, you said, pointing at it again. Inu. I told you it wasn't a dog, your favorite animal, but you did not believe me. Inu! you said again, hunching down on your hands and knees for a better look.

Luckily, the fox did not hear you; it would have been appalled to hear itself mistaken for its mortal enemy. Though kitsune have the power of assuming any shape and of making themselves invisible, the dog can always see its true form.

I tugged at your hand. The fox was closer than any I had ever seen, even in a zoo. I did not believe in its slumber, or that it would last.

I squatted beside you, our knees crunching into the wood chips, as you clapped with delight. Stay asleep, I told it. Let us watch you from a distance.

———

"In grappling with such a complex and commonplace figure [as the kitsune], one is quickly confronted with a troubling question of scope," writes scholar Michael Bathgate. "The fox—a symbol whose diverse meanings are interwoven with an equally diverse range of historical contexts—admits the study of any number of different phenomena at any number of different levels."[8]

As always, I feel I am failing in telling a true thing, a whole thing. What I know of kitsune is so little, refracted through so many lenses. The tale changes upon each telling. I try to expose all the different facets and yet I can't begin to talk about them all. So instead, I'll finish with my favorite: the kitsune no yomeiri, the fox wedding.

A fox wedding happens between two kitsune. It is not the same as when a human man takes a wife who happens to be a fox; those fox-wife stories always end in sorrow. In those stories the kitsune—always in the guise of a human woman—marry men and have children. In these tales they are loving and devoted family members until the very moment their true nature is found out, often by accident. At that moment of discovery—of rupture—they always return to the wild. They cannot stay in the human world.

(Whenever I write about being a bipolar wife, a bipolar mother, I receive emails from other bipolar women asking, *But who could love me?* How few of us are in community with each other. We do not have models to follow. So many of us only see media portrayals of people like us. How hard, then, to imagine the other paths possible.)

The fox weddings, instead, are celebrations. They occur on sun-shower days. When I was staying in a hostel in Nikko, in Tochigi Prefecture, a place whose lush greenery could not be contained by its cement sidewalks or brick walls, whose every sidewalk crack exploded in a riot of fiddlehead ferns, I experienced a sun-shower for the first time in my life. The kitsune are getting married, the hostel owner said, looking out the window at the rain sparkling in the sunlight.

Another way to know a fox wedding is happening is when you see a parade of ghostly lights: "At whatever time at night, whatever place, on occasions when it becomes extraordinarily quiet, flames like paper lanterns or torches can be seen, usually continuing far into the distance . . . This is what the young call 'the wedding of foxes.'"[9] All the kitsune gather for a fox wedding. They stop their trickery and return to the wild for the occasion. Unlike a marriage between a fox and a human, a marriage between foxes can endure. Each knows what the other is.

There are various theories for why kitsune choose to get married on sun-shower days. Some believe that the weather itself is uncanny, a form of trickery, like the kitsune themselves. Others believe that the kitsune cause the rain to fall so that humans will not venture into the mountains on wedding days.[10] (The top of the mountain, where the wedding will take place, will have clear skies, of course.) This gathering belongs to them.

This is why I love the story of the kitsune no yomeiri best: it is the one where they are not interrupted by humans. That is why we know so little about the fox wedding, because we are not there. They keep us away, knowing we cannot understand their customs; we cannot understand their love, or how they celebrate it. The way they laugh together, I think, is different from the laugh kitsune use in the folktales. That laugh is performative, that laughter is to scare. The way the kitsune laugh when they gather is only for each other.

SNAKE

The Snake, sixth in the twelve-year zodiac cycle,
symbolizes cunning, wisdom, jealousy, and fickleness.

SKIN, A LOVE STORY

———

Do you want to come over, I text him. *We could watch* The Sixth Sense.

Five minutes later, I send a follow-up text: *Is it weird that I'm asking you?*

The *him* here is Aaron, age twenty-eight, a recent graduate from the creative nonfiction MFA program I've recently begun. He is an adjunct. He plays Ultimate Frisbee. He sends me a video over Facebook Messenger of him chasing a goose by a pond. His style has been written up by a local undergraduate blog as *prankster prep.* He is kind and he makes me laugh and he is not, I think, serious.

I am twenty-two. I've spent the last four years of my life at a liberal arts college, population 1,400, in the middle of Wisconsin. It was the kind of school where, though everyone knew everyone's business and could count the number of Asian Americans on one hand, people still confused me with Cori, who matriculated three years after me. On the plus side, the school was only three hours from Chicago. Any time I started feeling *bad,* I could drive home and regroup.

Now I'm at a fifty-thousand-student R1 university in State College, Pennsylvania. Though it is, as my mother notes when she drops me off, the whitest town we've ever seen, it feels like a fresh start. For the first time in my life, I'm living outside the Midwest. I am, all things considered, well.

Here, in this place where no one knows me, I will be aloof and fancy-free. Maybe I'll have a floozy stage, I tell Cori.

Instead, I meet Aaron at an English department party the first week of grad school. A little while later, in a dark basement bar, he asks me if I want to dance. I always want to dance. The bar's beer-sticky floor makes each step feel like walking through mud, the exertion needed disproportionate to the distance gained. Still, I feel light as we spin around. I am twenty-two and I want to be different, and I want to date a million people, and I also want to be loved in exactly the way I want without ever having to explain anything.

But not explaining has its drawbacks. After a month with Aaron, I want clarity, or at least to know what to say to my friends when they ask.

One night we are lying on my bed, on top of the comforter, staring not at each other but at the ceiling, when I ask him what we are. The lack of eye contact makes the conversation possible. He is six years older than me and ten inches taller and lying down is the only place where I feel like I am on level footing.

What do you mean? he asks, his hands folded on his stomach.

We're spending all our time together, I say. He still does not understand what I mean. Through various fumbles I make it embarrassingly clear that I am asking him to define our relationship.

Oh, he says. Hmm, he says. I'd want to date, he says, if you're interested.

We are still not looking at each other.

I'm bad at dating, I tell him. I'm bad at relationships.

I don't say the word *bipolar*, though it hangs there in the air. He knows because he's seen my writing, and all my writing is about my illness. Earlier, when I asked him what he thought, he told me they were good essays. He only knows about my bipolar as something in the past. In the essays I control the narrative. In life, not always.

I think of my last boyfriend, the one I dated before the hospital, who I broke up with over MySpace the night before prom. The boy whose touch, by the end, I could still feel on the curve of my waist hours after he'd gone, both phantom and acidic.

I think of the boys I'd loved over AIM and LiveJournal and MySpace and Facebook. How many nights, how many boys, how many open tabs, how many times I mistook shared history for shared feeling.

When I say I am bad at dating I am not looking at Aaron, the man lying on the pillow next to me, our skin barely touching. I am looking at all those other boys still in the room with us, and all the girls who were, if earlier iterations, still me. After my bipolar diagnosis I'd imagined a transformation like a butterfly emerging from a cocoon—the way the books depicted it—but it was more like a snake shedding its skin. The skin underneath was new, but it still looked exactly like the old skin.

I tell him I will make a mess of things.

Okay, he says, and leaves. For a little while we do not hang out. We see each other only once, at a mutual friend's nineties party where he wears jeans with a jean jacket, and I am dressed head to toe in bright spandex. After the party ends, I go to a dance club with the other first-year graduate students, and he gets a drinking ticket for carrying what the cop writes up as a *one quart, eight oz beverage* while walking home.

———

I was born in the Year of the Snake. As a child, my ama told me this meant I was smart, though the place mats at Chinese restaurants told me this meant I was sneaky. In the Bible I read every night, snakes were bad—the *original* bad—and God had cursed them to slither on the ground. As a child I was ambivalent about being a snake. It did not seem as positive a symbol as my sister Cori's, whose monkey zodiac was invoked every time she climbed a tree.

But I did love to slither away from a situation, to burn a bridge behind me. Each school I left meant a group of friends to whom I had made promises and then broken them.

Of course, when I say I loved burning a bridge I mean I knew no other way. My approach to any interpersonal conflict was the same as the medieval approach toward infection: to let the wound fester until there's no choice except to chop off the limb. Unlike other members of the animal kingdom, we humans cannot regrow an arm or a leg. We cannot always start again.

———

Then it is October, the weather cold and the writing old and whatever resolve I possess falls along with the maple leaves outside. *Do you want to come over,* I text him. *We could watch* The Sixth Sense.

Meaning *I miss you,* though I could never say such a thing.

Is it weird that I'm asking you? I text him.

It is, a bit, he responds. And then ten seconds later: *I'll come.*

We watch the movie on my bed, where I also do all writing and grading and studying lying down. I'm teaching for the first time, trying to keep up with lit Ph.D. students after being a psychology major, and I think this level of fatigue must be normal.

I choose *The Sixth Sense* because it is October, spooky season, and because the movie seems free of any possible romantic subtext. I do not know what I want. I only know I want him here.

Halfway through the movie I pause and tell him that maybe we can try dating. He's planning on hiking the Appalachian Trail next spring anyway; I think the relationship will have a five-month trial run and a choreographed ending. I do not have to be the dumper or the dumpee. We can just fade out in different directions.

We tentatively start to use the word *us*. Until now, I have only ever thought of myself as singular: a me that would remain always alone and the center of the universe.

Our lives and daily routines meld together. When we go grocery shopping, we buy three-dollar crossword lottery scratch cards, scratching off each letter in turn. When we win five bucks, we forget to claim our money at Wegmans. I keep the old cards and use them to write a series of micro-essays about desire, with the constraint that I must use all the words on the card. I call it *Dreamscapes: The Scratch Card Series*. One is about how I want to have a child but do not think I will ever be stable enough to have a child. Another is printed in a literary journal that sends me fifty dollars—my first paid publication.

When I am sad or stressed, we walk to the TGI Fridays on North Atherton for discounted drinks at happy hour—beer for him, syrupy cocktails for me—or drive to Otto's for a big soft pretzel and beer cheese.

Still: we both think he is leaving.

Instead, he tears his ACL and his meniscus playing Ultimate Frisbee and needs surgery. I help him pick out purple pajama pants that are big enough to fit over his cast. I drive him to and from every single one of his classes. By the holidays it is clear he cannot hike any trail. He'll have to stay.

———

The snake, the ouroboros. "Half of this mythic being is dark and the other half light, as in the Chinese yin-yang symbol," writes

Juan Eduardo Cirlot.[1] In Japan snakes were seen as magical, since the shedding of their skin kept them eternally young and eternally terrifying.

The serpent yōkai known as uwabami is an enormous viper whose hunger knows no limits. It can swallow a grown man whole. In one story from the late eighth century, a general killed an uwabami after it ate too many people from the village of Nakamura.[2] Fearing the uwabami's wrath even after its death, the villagers built it a shrine where one thousand years later locals still offered snake-shaped tokens for protection. To prevent disaster, worship the thing that eats you.

In another story, a maid from a wealthy family mysteriously left her baby in the house every day to visit a nearby mountain. When she didn't return, the family took her baby to the foot of the mountain to find her. For a while, the mother would come down every day to feed her child, returning immediately after.

One day she did not appear. Instead, they heard her voice. I have turned into a serpent, she said, the words rippling like the stream that flowed down to the village. Now I hunger to eat my baby, so keep him away from me.[3]

The family vowed to raise the baby as their own. Some weeks later, when rain slashed hard and heavy on that mountain, the serpent mother glided down on a cascade of water. Once she reached the pool near her child's house, she transformed into her former shape—that of woman, that of mother. She let her child see her one last time. Then she sank into the water, washed away to the bottom of the sea.

These are the stories of snakes: hungry, abandoning. When someone calls you a *snake*, you know what it means.

In high school, when my boyfriend said, *Don't break up with me, I'm fragile*, he planted a seed of disgust that, overwatered, rotted until nothing could grow, not even hatred. To hate something is to care for it at least a little. And like I said: I loved leaving people behind.

———

In January, Aaron, now officially my boyfriend, is in bed next to me, doing physical therapy exercises with his resistance bands. Occasionally I help him. Push here, he says. Pull there.

By now I have fallen in love with him but I cannot tell him. I know my role: the one who needs more, wants more. (*To be bipolar is to be hungry, to be bipolar is to deny your hunger.*) My love, I know, is corrosive. When we hold hands, I squeeze his three times, *i - love - you*, hoping his fingers get the message.

In March he is still in bed with me. He can walk again; now it is I who cannot get up. Despite my weekly individual and group therapies and my cocktail of medications, the stress of grad school is overwhelming. One night I take pills again, another night I cut my skin for the first time in years. My thoughts drift to death.

The evening before I am supposed to turn in a big project for a seminar, my distress peaks. It is spring, my worst season. I am on the phone with my mother, sobbing, telling her I cannot go on, Aaron sitting on the bed beside me. My mother and I come up with a plan: in the morning I will go to the emergency room and check myself into their behavioral health unit—their psych ward. Tonight, I will watch movies to distract myself.

I'll stay tonight, Aaron mouths to me as I'm on the phone. I'll drop you off at the ER tomorrow.

When I hang up with my mother, I tell him, through a wall of tears, to leave me. Why would he stay when I'm like this?

Because I care about you, he says, and because I love you.

In response, I cry harder. It is the first time either of us has used the word.

The thing about my love is that it is, at times, manipulative. It is true that I tell him to leave because I think I am protecting him, and also true that I tell him to leave because I know he will not. I want reassurance and I do not know how to ask for it. I do not know how to say I am afraid, or how to form the word *stay*. Aaron is not tethered to me by blood. He has a choice. My greatest fear is that he will leave when I tell him he should.

This dynamic repeats over the next few years, in smaller and smaller ways, when I get mad at him for responding to what I say instead of what I think, for him not being able to read my mind. How long does it take until I feel like I can say what I mean? Three years? Five years?

The next morning, we hug in the emergency room parking lot and I tell him, finally, that I love him, too.

After I am admitted into the psych unit, my mother flies out. During visiting hours it is my mother and Aaron who sit on the edge of my bed, making small talk. To cheer me up, Aaron brings me a can of Easy Cheese that the nurse confiscates because the pressurized air is a danger. We eat it when I am released in a week. We celebrate my twenty-third birthday. We get married four years later.

————

I understand the outsize role women play in caregiving, but in our life, I am the caregive-ee.

My therapist tries to make me feel better when I tell her I feel guilty that Aaron does all the cooking.

Well, who does the grocery shopping? she asks.

He does, I say.

Well, who unpacks the food and puts it away? she asks.

He does, I say.

Oh, she says.

Long after we are married, I still struggle with feeling like I'm in debt. When I tell Aaron about this, he says: Well, you make my life better, don't you? We make each other's lives better.

I agree, though it does not seem equivalent.

Well, he says, even if you make my life *three* times better and I make your life *ten* times better, that doesn't change the fact that we are much better together, right?

And even though he is mostly joking, this is comforting. It acknowledges our reality: that because of my limitations, there is an imbalance to our labor, and there will probably always be an imbalance. And it acknowledges the truth: that our relationship is not predicated on balance, and not predicated on labor.

When other people try to convince me of my worth in our relationship, they try to make visible all the invisible labor I do. And of course, I forget my contributions: the paperwork, the taxes, the bills, our family's schedule, the garden, all the tasks that include few external stimuli and few choices. But that argument is still predicated on *doing*, on productivity, instead of on the fact that our worth—both in relationships and as individual people—is not determined by how much we do.

(My father struggles with this fact up until the moment he dies. I do not want the same for myself, though I keep circling the same question, wondering how little is too little, how much of enough is enough.)

———

Once upon a time, a white snake spirit named Bai Suzhen used alchemy and other magical arts to transform herself into a woman and enter the human world.[4] In this version of the famous Chinese legend, she falls in love with a man named Xu Xian and decides to stay among the humans.

The two marry and have a son and live their normal lives, and throughout the years Xu Xian has no idea what his wife truly is.

Their prying neighbors, on the other hand, recognize that Bai Suzhen behaves strangely. They begin to blame all misfortune—storms, loss of crops—on her presence. They know something is not quite right about her.

We can try to put on a face that others will recognize, but often the truth will out.

A local monk realizes who Bai Suzhen really is and decides to expose her. In some versions, the monk hates her because he is actually a tortoise spirit, jealous of her magical powers. In others, he is upset because Bai Suzhen is an accomplished healer, and the locals turn to her instead of to his religion. In still others, he hates her just for who he thinks she is: a monster.

No matter the reason, one day he brings her husband xiong huang wine and tells him that if he serves it to his wife during the Dragon Boat Festival, she will appear in her true form. Bewildered, Xu Xian follows the monk's instructions. After drinking the wine, his wife transforms back into a white snake right there before him. Xu Xian faints from shock.

The jealous monk captures Bai Suzhen and locks her in a tower sealed by thunder. For years she is trapped until her son grows up. This is the part of the story I like the most, because he saves his mother not by battling the thunder that keeps her trapped, or by defeating the jealous monk, but by winning first place in the national imperial exam that all students must take. By studying hard, and coming in first, he frees her. (One can see the appeal such a story would have for parents of high-school-age students.) The family is reunited.

The monk is forced to run away and hide in the belly of a crab.

This is a story, again, that I come to as an adult, a story I seek out. This story, which can be traced back to the ninth century, is one of China's four great folktales, and still widely known in Taiwan. Originally, though, it was a morality tale, a tale of horror. The monk was the protagonist, determined to save Xu Xian from the deceptive white snake spirit Bai Suzhen, who could suck out mortals' lives.[5] The story alluded to the dangers of sex and desire.

But as the legend was passed down, the humanity of the white snake was emphasized, her monstrous elements receded, and the maternal and romantic elements took center stage.

It is one of the best classical love stories, explains my Taiwanese translator Jenna, who introduces me to this tale. The people preferred a love story. They emphasized some bits, de-emphasized others.

But I do not know how to write a love story. We are not taught to map its caverns the way we are taught to trace our trauma. This old love does not titillate an audience the way my pain does. I have not perfected its narrative. Perhaps in this way it is new. This is the thing that shocks me every day: that I have this, a husband, a family, months and years together I keep track of like the Narcotics Anonymous coins I used to collect. I am an easy person to like and even an easy-ish person to love but I am a supremely difficult person to live with. Sometimes in the morning I wake up and think, Ugh—me again.

Is the honeymoon period over yet? my father asks me on the last trip the two of us will ever take together. By then Aaron and I have been married two years, together for seven. I laugh, stalling. My father thinks I do not want to answer, and he is right. Not because the honeymoon period is over. Because there never was one. The beginning of the relationship was the hardest part. Since then, it has only gotten easier. I was never starry-eyed. From the beginning it felt like the ordinary dailiness of love. I felt about Aaron the way old married couples do, with deep fondness and care and grumpiness and anxiety. What I would later learn is love.

———

Here is an inventory of the gifts my husband has given me over the years: An enormous velveteen pillow with arms and a back that functions like an armchair on my bed, where I still do most of my work. A long white body pillow that wraps around me on both sides. A hammock for the garden. A black narrow futon specifically for the kitchen floor, the place I often become overwhelmed.

We have set up our life so I can be horizontal in so many places, so I can live my life on the ground. We have a bed not only in our bedroom but in our family room as well. We have adapted, this is true; it is also true that for our life to function, he had to adapt more than me.

At night still I wonder: Am I only stable now because I have Aaron to keep me stable? Because I have my family? Is my wellness predicated on having bolsters to prop me up?

When I ask myself, *Is my wellness real?*, what I really am asking is, *Is it inherent?*, by which I mean *individual*. It is that American individualism, this false pick-yourself-up-by-your-bootstraps narrative that I get stuck in. The false narrative that tells me, stability that is interdependent, communal, and collective is less worthy—instead of something we all should be striving for. I know that instead of tunneling in on this anxiety, I should devote my energy to ensure other people can also access collective support, to support those in my community as much as they support me.

But sometimes tunneling feels so good. It is the motion my body knows. The people around me remind me to come up for air.

———

The story of us, of our love, isn't important because it's extraordinary. It's important because of how ordinary it is and for the many years I didn't think I'd ever be able to have an ordinary life. For many, to be bipolar is to think no one can love you. When I was twenty-two,

I did not think I could have a partner, a husband. I did not think I could have children. If you search *ouroboros* or *snake eating its tail* in my documents, you'll find nineteen different essays or stories where I mull on this image. I believed I was the self-destructive snake, the tautological snake. (In some versions of the white snake legend, when the son frees his mother, she then has to return to her spirit realm. She cannot stay with her human family.)

I thought to have the life I wanted I would have to transform into a different creature entirely and I knew that was impossible.

And yet: there we are in bed, watching this movie, watching the ghosts we know are there but cannot see, the ghosts that might be us. Here we are in bed, even still, even to this day.

TEN

The third part is the turn, the change,[1] the section that, as Yang Zhai says, "should aim to include something that surprises the reader."[2]

The third part is the one you get tripped up on. The third act provides a swerve. Sometimes the swerve contains conflict, but it is not conflict-dependent.

"However," Yang Zhai notes, "it is important to note that *ten* [the second part] and *shō* are like two sides of the same coin—they are not separate things. There needs to be consistency. The two verses respond to each other as well as move away from each other."

We are still in the same story.

But *ten* allows us to swerve. Divergent paths, same story. This is the benefit. The curse is that you might go where you were not expecting. It deviates. You thought you knew where we were going. We are going somewhere else.

DUGONG

—

The Okinawan dugongs carry the dead to Nirai
Kanai, the afterlife. Only a few remain.

THE TWILIGHT HOURS

Once upon a time, off the coast of Okinawa, there lived a creature called the zan—the dugong. Long ago the dugong had been the god of the sea,[1] but over the generations the sea had changed, and what the people worshipped changed with it. The dugong was demoted to carrying the gods' messages.

Later the Uchinanchu—the people of Okinawa—said that the dugong carried you to Nirai Kanai, the afterworld. Some believed Nirai Kanai was to the east and others said it exists at the bottom of the sea. No matter what the dugong's role—god, messenger, or escort—it was integral to the community.

Even so, the dugong was hunted for its meat, for its oil, for its flesh, which some stories said was an aphrodisiac and others said could make you immortal.[2]

There is a danger to capturing such sacred beasts. Legends warned the fishermen that if they caught and ate a dugong at home, those they loved would face disaster—their wives would die, or their family members would face accidents on the water. (Interestingly, the tales did not say, *Do not eat the dugong*; they said, *Make sure you eat the dugong at the beach, not at home*.)[3]

Carvings of butterflies made from dugong ribs have been found across the island.[4]

———

Once upon a time, my sister Cori tells my family, the dugong flourished; now its habitat is being destroyed by the construction of military bases. There are only a few dugongs left, despite all the ways the Uchinanchu have tried to preserve them and their land and seascape, despite all the ways the elders and communities have fought to stop the bases.[5]

Twenty percent of Okinawa is controlled by the United States, and now the military is trying to build a base in Henoko Bay—one of the world's most biodiverse areas—on soil containing remains of civilians who died in the Battle of Okinawa.[6]

It's not even good for construction, Cori says. All those bones make the ground unsteady.

She tells my family this tale as we stand beside the tide pools outside the Okinawa Churaumi Aquarium on the north side of the island. It is May 2017, the second month of my four-month fellowship to Japan. My parents and sisters have flown out to visit me. Inside the aquarium are walls of glass that stretch high above us, where we stand and watch creatures like the whale shark swim through the water. Here it seems like nothing separates us—not glass, not space, not time.

Outside the aquarium tide pools dot the shoreline. The five of us stand in a line and look out into the East China Sea, marveling at its clarity like the naïfs we are. It's my sisters' and my first time in Okinawa; our great-grandparents left one hundred and twelve years ago. We try to pronounce the words *Ryūkyū, Ryūkyū Kingdom*.

By the time we visit Okinawa, there are thirty-two American bases on the island and only a few Okinawan dugongs left.

But they still exist, I tell Cori, thinking this is comforting.

She dips her hand into the tide pool. I hate to say it, she says, but all these animals will eventually be a tale we tell to future generations. She gestures to the water beyond. At this point, we're just waiting.

That evening, after a stop at my great-great-grandfather's ancestral home, we return to our ryokan. Our usual boisterousness is diminished; we are exhausted from the journey. Before dinner I google which way to tie our yukata to avoid bad luck (the living tie the robes left over right, the dead right over left).

When my mother helps my father dress I see his body, which looks like mine. When we sit down to eat our meal, his hunger is a human's hunger, and when we dance together for a video clip before bed, his arms move the same as ours.

In Okinawa, my father is dying but we do not know it yet. In other words, he is not dying.

———

One month after my family's visit, a man in a local bar on the Shimane coast tells me I still have the genki spirit. *Genki* literally means "health," but the word is used colloquially the way in America we say *I'm good* when someone asks *How are you doing?* I'm good. I'm healthy.

By *genki spirit* the man means not only my energy as I tell him all the journeys I have taken and all the research I am doing—but the way I am telling it, my words overflowing. It is not hypomania; it is only that I have so much to say and few people to say it to, and this Japanese man is fluent in English, and willing to listen. He is surprised that I am twenty-eight. He assumed I must be younger to have so much enthusiasm. In my irregular updates to my extended family, I write, *I am doing well!*, always with an exclamation point.

When Aaron comes at the end of June for the last part of my trip, he tells me I look thinner than before. Are you eating? he asks. Yes, I insist, I just walk so much here.

Later I remember that occasionally I subsist only on onigiri from the FamilyMart. Still, I've managed to mostly take care of myself for the past three months without my normal medications and support system. For me, this is as genki as I can hope for.

I have dreams for this last month with Aaron. I have a list of all the things I want us to do: hike on the old post road, stay overnight at a temple, visit the Miyazaki museum, eat every kind of kakigōri, introduce him to my friend Mihoko. I still have research to complete, but he can come with me. Aaron and I pack up the tiny Tokyo apartment that has served as my home base, store my luggage, and take the shinkansen to Kyoto.

This is how to navigate the station, I tell him. This is how you buy a ticket. This is the best bento to order. It's taken me this long to understand what's happening around me, and now I am joyful with my tiny scraps of knowledge, my handful of words.

I have made it through three months in Japan without a crisis, and now my husband is here, and ahead of us I anticipate a month of excitement and ease.

———

As Aaron fiddles with the door lock at our Kyoto Airbnb I check my phone and see that my mother has left a message asking for a family video call. This, combined with the fact that my father had stomach pangs earlier in the week, feels ominous to me.

It's probably the CAT scan results, I explain to Aaron, as we lug our suitcases into the apartment. They take up half the space. The table is a sheet of plywood stacked on two milk crates. The lightbulb has no shade.

As we maneuver the bright red couch that unfolds like a piece of sandwich bread, I brood. A video call means bad news, I am sure, but I am also prepared. I've heard this story many times before.

In 2003, when I was fourteen, my father was admitted into the emergency room with abdominal pain. The doctors found a tumor in his belly; the surgeons cut it out; they told us it was not cancerous.

But over the next decade and a half he had six more hospitalizations, each one weeks long, due to accumulated intestinal scar tissue from the original removal.

Each hospitalization sent my father into psychological crisis. With his body confined for a month, his mind whirred and agitated. He had his laptop, he could access his patients' charts, he entertained visitors, but he wanted to be back at work. He thought he was useless; he feared he would never be useful again. He dreamed of wet-vaccing our carpets. My mother tried to talk to him about productivity and worthiness being different. It did not take.

The year before I went to Japan he'd had a six-week hospital stay, in which his despair was so great it broke through the barrier he maintained between his emotions and his daughters. Usually, only my mother saw the extent of his depression, how the pain could made him scream out, *I want to die, I want to die!* I'd only ever learned about these moments afterward, when my father was back home, and could laugh about it with us. I go mental every time I'm in the hospital, he said.

But that time when I entered his hospital room a week into his stay, my father was curled like a cocoon in the love seat next to the bed, crying quietly. I was alarmed. I had only seen my father cry once before, when his own father died.

You know, you're all old now, he told me. You're married. Cori has a job. Kristi's close to graduating. Maybe my work here is done, he said, his shoulders hunched. Maybe it's my time to go to the Elves.

I halted. It is *not* your time to go to the Elves, I told him, forcing cheer into my voice to cover my panic.

By the Elves, my father meant Valinor, the Undying Lands. He meant the place Frodo goes at the end of *The Return of the King* for his final rest. By the Elves, my father meant death.

I think it's my time to go to the Elves, he said again, softer this time, as if testing out the thought.

Dad, I said, the Elves do not want you yet. I made my voice firm, though I did not know what the Elves wanted. I only knew what I wanted.

Over the next few weeks, he perseverated on this idea. I thought that my having once been suicidal would make me a better bedside visitor for my father, who was temporarily, if not outright suicidal, then at least death-agnostic. It did not. When I saw him weeping, something in me destabilized. I changed the subject or gave advice or asked him to do cognitive behavioral exercises that had not worked on me.

Much of his psychological despair stemmed from his overwhelming physical agony, though he tried to last as long as possible without more intravenous opioids. He feared dependency and, like many doctors, was a bad patient. After a lifetime attending to the affliction of others, it is easy to compare your own distress and find it lacking. He could endure another ten minutes without pressing the button, he thought, another fifteen.

I did not want him to suffer, and also—I did not want to see his suffering.

———

It turned out that I was right: the Elves did not want my father yet. He was in pain, and he cried a lot, and his body healed slowly, and he was released. He recuperated to his normal self.

Now, a year later, I ask Aaron whether we should fly home immediately if he has to get another surgery. We push the couch into the final position.

You don't even know if that is what the call is about, he says, lying down to test out the couch, which is essentially two rectangular cushions with a large gap in the middle. His feet hang off the edge.

But it could be. I lie down next to him. The seam hits the exact center of my back.

You love to suffer twice, he says.

It is true: I am a paragon of premature pessimism, a doyenne of divining doom, a firm believer that simply imagining a nightmare provides its own protection.

And yet there is a limit to even my own speculations. When we finally have our family call later that day—my parents on the couch in their house, my sisters in their own apartments, Aaron and I in our tiny room in Kyoto—I am unwholly unprepared for the news that my father is dying.

My father is dying.

He holds my mother's hand and tells us, in the soft bedside-manner voice that makes his patients love him, that he had a CAT scan, and that the CAT scan revealed many tumors, and the tumors are cancer, and the cancer is terminal.

We find out he is dying on a video chat, while sitting in a room where our table is a piece of wood, and our bed is two cushions jammed together.

How long? Kristi asks. She is braver than I am.

One and a half to two years. My father says these words calmly, as if he is responding to the question *How long do these fishing lines last before they break?*

I have prepared for so many outcomes and yet somehow not this. Right up until the moment my father says the word *terminal*, I am certain he will live forever. Unlike my ama and her sister, whose immortality I believe in due to their supernatural octogenarian strength, it is his body's weaknesses—through which it has always endured—that provide this impression. Each time he is hospitalized only increases my belief that this is a chronic situation that will endure forever.

Here's the thing: death hides inside the familiar. That is its greatest trick. Yōkai hide and then show their faces where you least expect it. When my father tells us he is dying, my astonishment is total.

Over the call, my mother's weeping is silent. It is the three daughters whose howls swell and then roar, all of us at our own pitch and strength. Inside this tiny apartment, a gulf emerges between the world before and the world that exists now. An array of potential futures falls over like dominoes until there is only one left.

No one is immortal. Our grief ululates through the wires.

———

In Okinawan tradition the dugong serves as a psychopomp, a figure that guides souls to the afterlife, a role that functions both as guide and boundary. Mythology and religion possess so many boundaries between the living and the dead, as if without them, we will stumble where we should not go.

Some of these boundaries are spatial: in Greece there is the river Styx, in Japan there is the riverbank known as sai no kawara. Some are temporal: in Japan the spirits of the ancestors return specifically during the festival of Obon; in Taiwan, this occurs in the Ghost Month.

Spirits can also be bound to specific times, like the middle of the night or twilight, when the murky light hides as much as it uncovers. Twilight is called ōmagatoki, or "the hour of meeting evil spirits," the temporal marker that divides night and day.

When you feel a cold breeze rustle against the back of your legs. When you smell something peculiar with no referent ("like that of fish or blood," writes Matthew Meyer[7]). When you get the prickly feeling my mother calls the *heebie-jeebies*. When darkness falls upon you. These are the ways to know that in the twilight hours, a yōkai is approaching.

"Boundaries," says Komatsu Kazuhiko, "are where order meets disorder."[8]

But also: "We live in an era when conventional boundaries are continually disappearing."[9]

———

When I return home to Chicago from Japan, I become obsessed with time. Before I saw it as an expansive resource to be gambled with, and now I see it for what it is: something you can count on the fingers of a single hand.

We know we are in the twilight of my father's life, but we are unsure what that means. The timeline keeps changing, depending on which doctor and which scan. Two years becomes six months, after a scan shows dozens of tumors, and after a successful surgery, it's *a long time*, according to the nurse, though no one in the immediate family except me believes this.

My ama and her sister, on the other hand, think he will be cured. We are strong, they say of their spirits, making fists like Rosie the Riveter; I do not know how much has been passed down.

I do not know if he has the twilight of the poles, which in winter can last three weeks, or the twilight of the equator, which lasts less than half an hour. I tell myself I have no hope while actually squirreling it away for a rainy day.

While researching on the Internet, which provides little information on my father's rare cancer, I stumble across an article about anticipatory grief: the sorrow over the thing not yet happened. At first, I think, Yes—this is it—this is the name. But three paragraphs later, the flood of recognition is too much. I close the article; I do not want to know; I already know.

Living in the throes of anticipatory grief means being hypervigilant at every moment. To constantly brace your heart for the wound while still loving and being present with the dying. (*I do not want my father as myth / I want my father as father.*) How long can one live like this?

What I believe, in those months when he is dying, is that anticipatory grief must feel worse than actual grief. I do not know for sure—I have nothing to compare it to—I only know that fear and uncertainty create a gruesome whole that is more than the sum of its parts.

Later, after my father dies, I will learn that my intuition was correct. The weight of the thing to come was heavier than the weight of the thing that had happened.

———

In one Okinawan legend from long ago, some young fishermen capture a dugong in their nets and set her free. In return, the dugong tells them to go to the mountains, for a tsunami is coming. The fishermen spread the word far and wide, but only their own village believes their story and evacuates. It is easier to believe when we know the source, when we trust the mouth. The neighboring village stays home.

When the tsunami of 1771 hits—one of the worst in Japan's history— and the enormous tidal wave rolls in, the kind that reaches to the heavens, those who believed the dugong, who anticipated what was coming, survived. The rest died.[10]

Over the next two hundred years, Okinawa and the rest of the Ryūkyū Kingdom would have to survive wave after wave of brutality. In 1874, Japan's Meiji government colonized the islands and ended their sovereignty. In 1945, during the Battle of Okinawa, almost 150,000 Okinawans died or went missing, most of them civilians,[11] constituting between a third and a half of the prewar population. ("We became an event in a more powerful nation's history," writes Elizabeth Miki Brina in her memoir *Speak, Okinawa*, referring to this battle between the United States and Japan.[12] "A place for more powerful nations to attack and act against each other.")

Why would we expect either the Japanese or American military to care about the animals of Okinawa when they have historically cared so little about its people?

What can the dugong do against such things? Once the stories said they were the god of the sea, then merely the messenger. At one time, they could control tsunami; in later tales they could only warn of them. Now the dugong are on the brink of extinction.

(And still, the elders and the community fight for the dugong's life, for its home. They are still fighting now.)

Once upon a time Taiwan also had a sizable dugong population, presumed to be extinct since the 1990s.[13] I do not know when the Okinawan dugong will too pass into legend; certainly, it will be within my lifetime. I try to find out exactly how many are alive but it is impossible to ascertain. We hear three, we hear two. There were around fifty in 2003.[14] In 2019, the International Union for Conservation of Nature cited the population as ten or fewer (along with, again, showing the threat to the species by the construction of the U.S. military base at Henoko Bay).

The Defense Ministry in Okinawa confirmed three in 2018, but has since found one dead, and no sign of the other two.[15]

I hope they are hiding far away from the military bases, under thick beds of seagrass.

Fewer than ten, two or three—how much easier it is to confirm death than life. How hard it is to know how long twilight will last when we do not know what season it is, where we are on the longitude and latitude of the earth. All we know is that we are facing the dark.

———

While I'm waiting for my father to die, I try to work on my novel. Only I have nothing to say about my characters, who live along the sea and make their living fishing, and whose journey is triggered when the decapitated head of an oni washes up on the shore—its bloated red head and horns foretelling all the danger about to arrive on their island. I went all the way to Japan to write this book and now I cannot write it.

Or rather: I *do* write, thousands of words a day, but the story does not go anywhere. I do not know the ending, or how to get there. Having lost the plot, I aim to fill a quota.

I read more about the four-act structure originating in Chinese poetry that in Japan is called kishōtenketsu. This structure hinges not on conflict, but on complication. The first act is called Ki— the Introduction. This act introduces the reader to the world of the story and the life of its characters. The second is called Shō—the Development. The story deepens; the reader develops an emotional attachment to the characters. The third is Ten—the Twist or Complication. This act is usually unrelated to the previous two. Something unexpected unfolds. And the fourth is Ketsu— Reconciliation. This act concerns the effect of the third act on the previous two.

The structure is fascinating to me, and yet I do not try to apply it to my novel. Instead, I write scene after scene of heroes fighting yōkai, of burgeoning teenage love, of magical showdowns, conflict after conflict, and each one sounds like people fighting with cardboard swords. *Swish swish.*

How is the book going? people ask. It's good, I say. How are you doing? people ask. I do not know how to respond.

What is there to say? The dugong cannot stop the tsunami: it can only tell you it is coming.

HYAKKI YAGYŌ

———

The Night Parade of One Hundred Demons, the procession
of yōkai that cavorts down the streets on summer nights.

THE NIGHT PARADE

—

SPRING 2018

This is an ongoing tale, the ending uncertain. In the morning I examine my toilet paper. At night I read my underwear like tea leaves, looking for the rust that means ruin. I tell myself: This is okay, and this is okay. Today, everything is okay. Once in a blue moon I believe my own stories.

My pregnancies—the one I lost and the one that is still yet to be determined—I can only bear in fragments. Thus, the retelling is fragmented. Perhaps this is a convenient excuse, a way to explain the lack of narrative cohesion, a way to sidestep the niggling issue that my professor once called the "so-what" of the essay.

I lost a baby. *So what?* As of today, I am having another. *So what?*

SEPTEMBER 2017

These days I can't stop thinking about origins. Maybe it's the time of year: autumn in Chicago, the briefest of seasons. Or maybe it's because September is the month of Rosh Hashanah, one of our three annual New Year's celebrations, and the month Aaron goes back to work.

Or maybe it's because this month my doctor removes my intrauterine device. We are *going to start to try*, a vague, euphemistic phrase that piles beginning upon beginning.

Originally (*in the beginning*) we planned on waiting until the following year to remove the IUD. But then my father, his diagnosis. My generous, loving father, who, dopey on drugs after his debulking surgery, tells me how excited he is for grandchildren. He knows we want kids. He wants to be called Agon, the word we used for his own father, who swung me in a blanket and died when I was young. At church, when he sees my friend's toddlers poke their heads out of a pew in front of us, he leans over and whispers, I want one of those.

So: the gynecologist's office. Here my legs latched into stirrups. Here the pinch as the doctor pulls something from my uterus. Here in her latexed palm the only thing I have to give, the copper coil of time.

I cannot think about where my father will be in nine months. Even as I type this it hurts. So, this is enough.

September is also the month I join a new Bible study. My father is dying; I need something to believe in. We are reading through the book of Genesis. The majority of the tales in the first part revolve around the men—Adam and his rib, Noah and his ark, Abraham and his covenant, Joseph and his dreams—but I am more interested in the matriarchs. I understand their longing. I understand their need.

For example: Abraham's wife, Sarah, who had yearned for a child for years, was ninety when God told her she would finally—*finally!*—bear a son. She laughed in God's face.

He was displeased with her incredulity, asked: *Is anything too hard for the Lord?*[1]

Sarah was afraid. *I did not laugh,* she told him. But he knew.

I understand her fear. Her disbelief that God would give her a good thing.

———

After the IUD removal I whip myself into a frenzy of action. I consult with my psychiatrist about how we will taper off my medications once I'm pregnant. I consult with my therapist about how we'll deal with all my anxiety.

But for every minute I spend trying to calm myself, I spend sixty perusing the baby boards, the forums that serve as the meat and potatoes of pregnancy and childrearing websites. The articles by professionals do not have the allure of the forums. There's a post by a doctor that says spotting and cramping in early pregnancy can be completely normal; there are also thousands of excruciatingly detailed replies by women for whom spotting was the first sign of the end. The Internet Age: the information is endless, and endlessly terrifying.

From the boards I learn about how these women, many of whom have experienced infertility or repeat pregnancy loss, look for symptoms—sore breasts, nausea, extreme hunger, lack of hunger—during the two-week wait between ovulation and a pregnancy test. Most of these women admit up front that what they are doing is illogical, that the signs don't really mean anything at this point. And yet here they are. Here I am.

It's ten in the morning and I'm on the boards. It's midnight and I'm on the boards, the only place where the tenor of the discourse matches the dialogue inside my head.

On the forums the women understand without needing explanation. They do not try to dissuade you from your irrational thoughts. They do not spew facts in the face of your distress. Instead, they say, *Hugs, mama*. They say, *It's so hard*. They say, *I know*. They know the fear that waits and watches, that breathes every time you take a breath.

I grew up knowing that wanting a baby in no way guaranteed getting a baby. The women on both sides of my family fought for the children they had. The miscarriages, the infertility. The cousin, born on my birthday, who died two weeks later from heart defects. My aunt and uncle praying at night with a different cousin for years: *And please bring him a little baby brother*. My mother probably had the easiest time of all; still, we grew up with a rosebush in our garden to honor her miscarriage.

People in my own life point out that I have not experienced any loss or pregnancy-related struggle of my own. They marvel at my well-being. I am twenty-eight. I have not had a bipolar episode in five years. I have been with Aaron for six.

But what looks like health is a stability accrued with effort, one hour at a time. How long I spent thinking I could never sustain a long-term relationship, how I could never be a mother. They do not understand how tenuous it can feel, like living on a fault line.

Even Aaron, who sees so much of me, struggles to grasp the depth of my fears. He is a math teacher who deals in logic and measurements. He counts months: If we get pregnant now, then the baby will arrive this month. If we get pregnant in this month, then the baby will arrive in *that* month. His steady reason is a big part of why I love him, and yet he does not realize how each of my worries is a frayed thread that when tugged will collapse the entire bolt of fabric.

I've spent so long discussing motherhood with my therapist and psychiatrist, both of whom believe, as I must, that with the help of my support system I can do this. It's just that I do not know anyone who *has*—done it successfully, I mean. I do not know any other bipolar parents. Such figures appear in novels, or in memoirs written by their children, and these stories are filled with hospitalizations and trauma and chaos caused by unmanaged conditions. I have never read a story by a mother whose disorder is managed.

All I want is one, one single tale, telling me it is possible. Difficult, yes, but possible.

OCTOBER 2017

Our first month trying. During the two-week wait, I'm convinced I'm pregnant. I am closely attuned to my body in the way of the chronically ill. *I can't tell anyone because they'd think I'm crazy*, I write in my journal. *But I'm writing it here, first, as proof.* I take a test much too early and wait for the second line to appear. It does not. Next day: same routine, same result.

Three or four days later I start to spot. I tell myself it's only implantation bleeding—the result of an embryo attaching to your uterine lining—but the next day's gush of blood proves otherwise.

Each month I do not get pregnant is a significant percentage of my father's remaining life span. In my chest, a daub of sadness that will soon be sublimated by other, larger griefs.

I pace. I have ample time for pacing; I had to quit my library job to go to Japan. My bosses tell me they'll hire me back as soon as there's an opening—probably within a few months—so I half-heartedly apply for other positions while waiting for them to call.

At my coworking space in Logan Square, where I answer the door and refill coffee in exchange for a free membership, I type and type and type. I make sprawling family trees for my characters, ten generations deep, like those Old Testament lineages.

But I am not like the biblical matriarchs. They may doubt, but they always come around in the end. I do not see my face in their faces—

waiting, obedient. Mine is mirrored in the woodblock prints of yōkai in the folklore books I'm reading. (If it's for research, I reason, it counts as writing.)

During summer nights, the stories go, all the yōkai journey through the streets of Kyoto, the old Japanese capital, accompanied by flags and music and general conviviality. Here, the fluffy tanuki with the huge testicles bats his fan at the ittan-momen, whose bolt of cloth whips in the wind. The old yamauba cavorts with the kodama and the tengu.[2] This procession is the Hyakki Yagyō, the Night Parade of One Hundred Demons.

The villagers know to hide when the Night Parade comes, for anyone who has the bad luck to stumble upon these yōkai will be spirited away. But for the yōkai themselves, this event is as boisterous and joyful as any summer matsuri. In the Night Parade, they are among their own kind.

Some of these yōkai are women, and these women look more like me than not.

For example: the futakuchi-onna.[3] Look at her neat kimono, her polished face. You would never know she is a yōkai. The only way to tell her presence is by absence: the way food around her disappears at alarming rates.

If you lift her hair—hair long and thick and dark, the kind I've been complimented on my entire life, hair that can be pulled across the face like a curtain under which to hide—if you lift her hair you'll find, at the nape of her neck, a second mouth.

An old tale: a woman turned into a futakuchi-onna when her husband, chopping wood, split the back of her head with an ax. An accident, presumably, though the woman had earlier starved his child—her stepchild—to death, so you never know. In any case, the wife didn't die. Instead, the gash at the back of her head transformed, as old wounds are wont to do. Scalp into tongue. Occipital bone into gnashing teeth.[4]

Look at those teeth, those lips, that ravenous jaw. Into that mouth all the food, all the everything. The futakuchi-onna can use her hair like tentacles, grabbing all the food in sight. She let her child go hungry, and now she herself will never be sated.

NOVEMBER 2017

Our second month trying. Though I use ovulation predictor kits and take my temperature to track my cycle's thermal shifts, I'm sure we didn't time our sex correctly, and that I'm not pregnant. Unlike September, when I started peeing on a stick a week before my period was supposed to arrive, I don't think about it at all until I realize my period is late. The test shows me two bright blue lines.

I call my husband's name from our tiny bathroom, opening the door with my foot.

Aaron, hearing something in my voice, strides over—the fastest I have ever seen him move when not exercising—and stares at the test I hold in two hands like an offering, my underwear still at my ankles.

Hurray! he says. And then one beat later: Can we not have a gender reveal party?

We tell everyone very early. My parents and sisters are overjoyed with the news. My father is delighted to join the ranks of his friends who are grandfathers. They can't have all the fun, he says.

My psychiatrist tapers me off my medications. My mother's cousin texts me: *You think it's your baby, but it's all of ours.* I am the oldest in my generation on both sides of the family, and they are all deeply invested in the life growing in my womb. This is what gives me peace when I think about being a mother with bipolar disorder, when I think about raising a child. We will not be alone.

DECEMBER 2017

At the first ultrasound we see the gray blob that will become our child. The crown-to-rump length marks the age at eight weeks, one day. I take a picture of the sonogram and send it to all the family group chats. Afterward, we go to the Stan's Donuts on Erie Street to celebrate. Aaron gets a coffee. I order a cruller, a Boston cream, and an apple fritter. I am pregnant.

When I start spotting the following Sunday, I remind myself that this can be normal. There's no cramping, not enough blood to fill a pad.

Three nights later, when the blood keeps coming, I message my midwife. *It might be nothing*, she responds the following morning. *You can wait and see. Or, if you want, we could do an ultrasound, just to check.*

I do want. I always want. I call her office and take the ten o'clock appointment. It is 9:25. If I leave immediately, I might make it.

I drive, frenzied, sobbing from my gut. I leave a message for Aaron at school. I call my mother, who says she will drive in from the suburbs to meet me. *Please, please*, I pray as traffic stalls along Lake Shore Drive. Out my window I see a few people walking their dogs along the beach, undeterred by the cold. *Please let the baby be alive, please let me get there in time.* At this moment both these requests seem equally important. I'm probably overreacting and yet my body cannot stop dry heaving. I feel deranged.

By the time I find parking I am late. It doesn't matter; the office is running behind. A nurse hands me a clear plastic cup in which to pee. It comes out clotted with red strands. Finally, I am led into an examination room. My mother arrives right as my midwife removes the ultrasound wand. Her silence is everything. I am supposed to be almost ten weeks. The fetus measures eight weeks, four days.

Another doctor confirms there is no heartbeat, no blood flow.

I'm sorry, she tells me.

I'm sorry, I tell my mother, who is weeping harder than I am. My own tears are minimal now, unlike in the car, when I still didn't know. Now I know.

My mother and I hold hands. I am glazed.

Because I am so far along, the midwife recommends a D&C. Aaron takes off work, and we go to the hospital the next day.

Make sure you ask for extra blankets, says my father, who as both doctor and recurrent patient is a connoisseur of hospitals. They have a machine to warm them up, he says, it'll keep you toasty.

For many hours Aaron and I wait in a small room, dark and blessedly private. I wear a paper gown. I ask for the extra blankets. It is all simple and painless; the worst part is when the nurse can't find my vein. I am put under. They take everything dead out of me. When I wake, we go to Shake Shack, and I order a 'shroom burger and a chocolate malt. We take a photo and send it to my family to show them I am okay.

I feel that I have let my entire family down. My mother disseminates the news. When my grandparents call, when my aunt leaves a message, I do not pick up.

————

Over the next month I read a lot of articles and blogs about miscarriage. All the top Google hits are by wealthy white women whose posts conform to specific aesthetics, the text juxtaposed with well-lit photographs of the writers looking into the distance or rocking in a wicker chair with the sun setting behind them. These posts are usually one-offs, a single, intimate glimpse on a blog dedicated to organic gardening or DIY furniture refurbishing. What happens after? The posts do not say. Miscarriage is treated like an event that happens once and is left behind. I prefer reading the infertility blogs, where grief infiltrates even the most mundane and unrelated posts.

I find nothing that truly resonates—no quote that I can watercolor with my sadness and hang on my wall.

———

Two weeks after my D&C, Cori sends me a submissions call for an anthology about miscarriage and infant loss by women of color. It is exactly the thing I wish I could read right now.

Send something! she emails. I don't how to write about the miscarriage without writing about my father, how to write about both of those things at all, or how to begin when there are no more beginnings left.

———

For example: the hone-onna.[5] Woman of bone. Woman who used to be human. Who may not know she is dead. Woman whose love keeps her spirit alive. Whose love keeps her appearance the same as in life. Who will inevitably return to her former lover. Who will inevitably return to her former lover's bed.

(You might think: Wouldn't he know the woman he loved is dead? Yes, but do not overlook how much grief thirsts to be deceived. How much we desire to have a veil drawn over our eyes.)

The couple may stay together a day, a week, a month. The couple who is delighted at being reunited. Who cannot believe their luck? Don't forget that in this kind of story there is always a price to pay, and like in all these stories, it is the living who pay for the dead.

Watch the hone-onna's lover wither away as his life force disappears. It is she, of course, who is sucking him dry. The dead cannot survive on nothing. If they stay together long, he, too, will die. (Is this what she wants? Is this what he wants?)

The only people who can see the hone-onna for who she really is—a rotted, wasted woman of bone—are the people for whom love does not have a hold.

But who wants to see with such true eyes?

JANUARY 2018

My midwife tells me to wait one cycle before trying again so my body can heal. So we wait. For the first time in our marriage, we use condoms, preventing the thing we most want.

On January 10, after reading an article that appeared in my newsfeed, I call my father with panicked medical questions about the Zika virus. He answers each one, his voice, as always, a balm. Hours later my mother sends a GIF of balloons and party hats in the family thread, and I remember it's his birthday. I call him back to apologize and send my well-wishes. No problem, he says cheerfully. How many times, over the years, have I forgotten his birthday until late in the day? How many ways can I tell him I'm sorry?

This month I find a quote worthy of watercolor, only it's not in an essay about miscarriage. It's not about miscarriage at all. Early in Hanya Yanagihara's *A Little Life*—a book so filled with injuries that I eventually cannot finish it—a father ruminates about his child's death:

> But here's what no one says—when it's your child, a part
> of you, a very tiny but nonetheless unignorable part of
> you, also feels relief. Because finally, the moment you have
> been expecting, been dreading, been preparing yourself
> for since the day you became a parent, has come.
> Ah, you tell yourself, it's arrived. Here it is.
> And after that, you have nothing to fear again.[6]

I would never compare my first-term pregnancy loss with the death of the character's son, but I recognize the impulse. I am devastated, yes, and I had expected to be devastated.

Forty days after the D&C, I get my period. I start tracking my temperature again, and in another three weeks I reach my fertile window. We have a lot of sex.

While we wait, I begin to look at the yōkai I researched for my set-aside novel with new eyes. I think of the book I wish I could read right now. I do not know how to write about loss yet, but I know how to write about monsters. Whether I want them to or not, the yōkai always lurk nearby.

FEBRUARY 2018

I tell Aaron I won't take a pregnancy test until I miss my period. He tells me that's a good idea. It *is* a good idea.

I test six days before I'm supposed to miss my period, the earliest you possibly can.

Two lines appear. One bold—the control line—and one the faintest apparition, a mere shadow of blue. I text a photo to my sisters, to a few of my close friends. *Can you see this? Is it there? Is it real?*

They reassure me that they do, indeed, see something, though what it is I cannot yet say. More data is required. Day after day I test and test, watching the line grow darker. My period does not arrive.

For a little while I am elated. It feels like a miracle. After months on the baby boards, I know how long women wait for what they call their rainbow babies, the ones that come after the storm of miscarriage, stillbirth, or infant loss.

But the thing about fear is its insidiousness. A drop can taint a whole sea of happiness.

What they didn't tell me to expect when I was expecting again: how it feels worse in some ways than those weeks immediately following the miscarriage, when I wasn't yet pregnant. Then I was afraid in the abstract. Now that I am again *with child*, I carry the fear deep in my belly.

When a good thing has been taken from you, it is hard to believe that you will be able to hold it in your hands.

Do not be anxious about anything, the Bible says, *but in prayer and petition, in every situation, with Thanksgiving, present your requests to God.*[7]

I present my requests. And still, I am afraid. If only I were stronger, I think, if only I had more faith.

MARCH 2018

I cannot talk about the pregnancy without qualifying it: *if, if.*

My psychiatrist starts to taper me off my medication. Since the first time went so well, I am not expecting any problems. I am wrong.

The thing growing inside me is eating me alive synapse by synapse, spreading like a virus into my limbs and fingers. All I want to do is break things. In the decade since being prescribed anticonvulsant mood stabilizers, I've forgotten what this psychomotor agitation feels like. Unlike before, the restlessness arrives without any trigger.

At night I cannot sleep. During the day I cannot stay awake. Aaron cooks me breakfasts of roasted garlic and eggs and mini bell peppers, and I worry about all the ways I must be wearying him.

My psychiatrist says my agitation is probably triggered by withdrawal. He adjusts my dosing schedule, so I wean off more slowly. My therapist suggests ripping up the cardboard boxes from our recycling. Ripping releases my energy, leaving my fear.

On occasions when the cardboard isn't enough, Aaron tucks me under my twenty-pound weighted blanket and climbs on top of me, curving his spine to allow space for my belly. He grabs me with his arms and legs and rocks me back and forth. Squeeze me as hard as you can, I tell him. Harder than that. Harder.

When all else fails I stomp around the house. I am angry at my psychiatrist, who told me I would be fine, angry at a medical industry that prioritizes the theoretical, infinitesimal chance of my medication harming a fetus over the concrete harmful effects of lack of medication on a mother. I am angry at the midwife, who tells me she doesn't know what to do. I am angry at myself for not voicing my concerns earlier, for not asking my doctors at the beginning if we had a backup plan.

These episodes last only fifteen minutes and yet they feel like forever. Afterward, a half hour of me crying, of me apologizing to Aaron, I'm sorry, I'm so sorry, what grown woman, what kind of grown woman—

The rest of the time I am exhausted. I write in three-word sentences. I am too tired for subordinate clauses and complex syntax, too tired to focus on more than one paragraph at a time. I cannot figure out what the characters in my novel are supposed to do, or how to make them do it.

At my weekly Bible study I stay mostly quiet. When the leader prays for me, she calls me by the name of the other Japanese American person in the group.

Though the Bible says that faith is being sure of what we hope for and certain of what we do not see, I am soothed by the tangible: heartbeat, Doppler, ultrasound.

What does it mean to trust, I write in my journal. I carry my fear like
the baby's invisible twin, or more like a subchorionic hematoma,
a condition in which a pocket of blood forms in the uterus. The
hematoma can shrink into nothing or cause a miscarriage; if you see
this on your ultrasound, it's a waiting game.

I know that expecting the worst is not an effective weapon against
calamity, but it seems like all I have. A balm that poisons.

My ultrasound occurs at seven-thirty at night. The hospital is empty,
the room large and dimly lit, barren but for the table I lie on and
the ultrasound equipment. The technician inserts a transvaginal
ultrasound wand inside me and moves it around, taking image after
image. It feels like a hospital out of a horror movie.

The process takes over thirty minutes, in which the tech is very silent,
and my mind is very active. I cannot interpret the gray splotches that
whorl on the screen, as cryptic as Rorschach blots.

After I pull up my pants, I scan the technician's face for signs. You'll
have to wait until your midwife calls you for your results, she says.

On the ride home I tell Aaron that I'm sure her abrupt manner means
something is wrong.

Aren't the techs not allowed to tell us any information? he says.

Didn't she seem *extra* distant? I press again. He shakes his head; he
didn't notice anything. But he wouldn't. He sees an ultrasound wand as
just a wand, not as Chekhov's gun.

In the days until my midwife calls me, I examine the photograph of our
ultrasound and imagine all the potential problems. *Please*, I pray, *please
let it only be a subchorionic hematoma.* In that case we still have a chance.
A hematoma, I can handle.

———

When Rachel saw that she was not bearing Jacob
any children, she became jealous of her sister. So she
said to Jacob, Give me children, or I'll die!

Jacob became angry with her and said, Am I in the place of
God, who has kept you from having children?

Then she said, Here is Bilhah, my servant. Sleep with her so
that she can bear children for me and I too can build a family
through her.[8]

In other words, we make our own contingency plans.

———

The midwife calls. Everything looks fine. Nothing is wrong. I did not
think of this as a real possibility.

———

For example: the zashiki warashi.[9] Like a poltergeist, the zashiki
warashi, which appears as a little child in a red coat, is tricksy but not
evil. If you're young, you might be able to see one out of the corner
of your eye. Having a zashiki warashi live in your home is good luck,
though your prosperity will fade if it decides to leave.

Some believe the zashiki warashi are the spirits of babies who died
early, whose bodies were kept inside the home, who were not given
the full funeral rites, being so young. Their spirits remain with their
remains. Protecting the home they would have lived in, a future no
longer theirs.

APRIL 2018

Despite my own anxieties, the pregnancy progresses. One night Aaron
and I pick up shawarma from Taste of Lebanon when Cori and her
then-boyfriend Sherick come into town. Nausea strikes as soon as

we climb back in the car to go home. I puke into a flimsy cellophane envelope—hurriedly removed from one of Cori's art prints—while she and Sherick cover their eyes in the back seat. Afterward, as I slowly drive us home, Aaron holds the bag of vomit between his finger and thumb.

The car fogged up like in that scene with the velociraptors in *Jurassic Park*, Aaron recounts later to my parents.

You have a good husband, my parents tell me.

At the end of the month the nausea disappears. I know that pregnancy symptoms tend to wane when the placenta takes over. I am relieved not to be nauseous all the time, and yet in its absence I do not find peace. I had no symptoms with my first pregnancy, and that one ended; some studies have shown a correlation between lack of morning sickness and miscarriage.

Jami, my husband says, with the mix of love and exasperation so common to marriage, there is no pleasing you.

———

My father's most recent CAT scan shows eight new tumors. We are not sure if we can go on our trip to Taiwan this summer.

On the plus side, he tells me—when I call just to hear his voice—on the plus side, if I get the surgery in the summer, it means I'll definitely be out by November, and we won't be in the hospital at the same time!

That had been one of his worries, that he would be recuperating from his operation at his hospital in the suburbs at the same time I'm delivering in the city.

My parents speak about the baby like it is a given. *When* you have the baby. It is so hard for me to believe this, though it is the most statistically likely scenario.

I start to worry about all my worry. The Internet tells me it's bad for me, that stress can lead to miscarriage. So I talk to the baby. I sing to it. When I hear the heartbeat at my ten-week appointment, I begin, slowly, to believe. I create a Pinterest board of all the crafts I want to make for the nursery. I start knitting a Totoro stuffed animal.

———

We announce our pregnancy at twelve weeks exactly. The night before, curled in bed on my left side—the better for the baby to receive blood flow—I envision different versions of status updates. I want to talk about my miscarriage, but it seems wrong to mention that in the same breath as a new baby. I worry people will not know how to respond, that they will feel compelled to temper their happiness the way mine is tempered.

What I want to say: *Today we are twelve weeks. After everything, getting to this point seems like—is—a miracle.* What I say: *The Center for Jewish-Japanese Relations* (Aaron's and my joke name for our household) *is premiering its best project yet!*

What I want to say: *A lot of fear. A lot of trepidation. And yet: a lot of prayer and a lot of joy.* What I say: two star-eyes emoji, two puke-face emoji.

After posting this truth-but-not-the-whole-truth announcement, I worry about how it will affect an old college friend who has written publicly about how hard it is to see pregnancy announcements after her own infertility and loss.

Among all the likes and congratulatory messages, her comment—no text, only an image of a rainbow—stands out to me the most. Such grace.

———

For example: the jami, or evil spirit. The yōkai that bears my name is the "ominous manifestation of uncanny malevolence."[10] In the illustration from Toriyama Sekien's encyclopedia, the jami looks a little bit like a lion dog, though it is floating in the air, and the lines that denote its fur drift up and up and up, until they become one with the wind. One theory posits that the name relates to the verb *jamiru*, which means "something going wrong midway through."[11]

For example: nurarihyon, the commander of the yōkai. Except nurarihyon is not a woman and he is not young. (That, of course, is why he is the commander of all the yōkai.) He is wrinkly and his head is shaped like a gourd. As commander, he leads the Night Parade. The other yōkai follow him down the streets. He is the elder, though his leadership is disputed.[12]

When he's not leading the parades, the nurarihyon is probably sneaking into your house. He chooses the moment when your entire household is in a tizzy, preparing for this celebration or that event, and then he shows up in his palanquin and lets himself in the front door. He'll act like he owns the place. While you're bagging up your period trash or scrubbing the sink with a Magic Eraser, he'll come in and drink the good Taiwanese tea. He'll use the fancy plates and eat the cookie assortment you were saving for guests.

Eventually you'll notice that there is a small potato-looking man in a silk kimono kicking back on your couch, crumbs everywhere. His power is so strong that even though you'll notice, you'll just keep on cleaning.

Only after he's left in his palanquin do you think: What?

What power—to make you see without really seeing, know without really knowing. To only discover in retrospect—and even then, not fully.

———

Another of my anxieties: Who am I to speak of this sorrow, when so many other women have seen much more? My relative luck is most salient to me on the baby boards, where you have the option of adding a signature, the text that appears every time you post. Instead of listing a phone number and address, as you would in an email signature, people list their births and losses in a type of shorthand.

DD 3/99, ^DS^ 6/02 28w, mmc 10/04, one might say. This woman had a "darling daughter" in 1999, then lost a son in 2002 at twenty-eight weeks (on the baby boards, carrots around a name indicate the baby has passed away; the symbols look like they're pointing above), and a missed miscarriage in 2004. Or *ttc #1 since 2012. 6 mcs, PCOS*. This woman has been trying to conceive her first baby since 2012. She's had six miscarriages and suffers from polycystic ovarian syndrome, a condition common in women struggling to get pregnant.

Many of these people would love to be in my situation. Who am I to tell this tale?

But also: There is no *this* tale. The story of miscarriage is not singular. It is plural, it is various, and it is frequently silent.

The *so what* of my story is that I am afraid. I am afraid of losing this baby. I am afraid of having this baby and losing my father. I am afraid of losing them both. I am afraid of losing my mind, which some days seems like it is hanging on by the most tenuous of threads. I am afraid of postpartum depression and postpartum psychosis.

I become more and more attached to the idea of this baby.

———

After I post on social media about being pregnant and bipolar, a writer I know reveals to me that she is also a bipolar mother. She tells me that she's thriving, that she loves being a mother. Her child is joyful and flourishing. I am so grateful for this one story.

Right after I got pregnant for the second time, I told my primary care physician how afraid I was. That's common in early pregnancy, he said. He paused, then added: I hear it usually goes away by the time the baby is eighteen.

I tried to make the fear leave. It wouldn't. Now I just try to stand alongside it, to keep standing. It kneels when I pray. It whirls while I fret. It watches over my shoulder as I write. It picks through the stitches while I knit a circular baby blanket, which, when finished, will look like a chunk of coral or the ocean.

I tell people about the pregnancy. I speak the news into the world like an incantation, as if the more times I say it, the more real it will become.

———

For example: the villagers who, upon seeing the Night Parade enter their village, would close their shutters. To see the Hyakki Yagyō meant death. Inside they would shut their eyes and murmur a string of sounds, an invocation that has no direct translation: かたしは , えかせにくりに, ためるさけ, てえひ, あしえひ, われしこにけり! かたしは , えかせにくりに, ためるさけ, てえひ, あしえひ, われしこにけり![13] This chant, they knew, would keep them safe. This chant would make the Hyakki Yagyō pass by their door.

So: I am having a baby.

Right now, this is true.

YŪREI

───

The faint spirits who are stuck on this earth,
unable to cross over to the afterlife.

THE HOUR OF THE OX

—

I

In America, the middle of the night is called "the witching hour." We all know what happens during the witching hour: the monsters emerge, werewolves howl, sorcerers cast their spells in the moonlight.

In Japan, it's called "the hour of the ox." We all know what happens during the hour of the ox: the fabric between the living and the dead thins to permeability. The yōkai prowl. The spirits known as yūrei cavort, haunt, keen, and avenge.

The hour of the ox is actually two hours, between one and three in the morning; in ancient Japan, as in other East Asian countries, each day was divided into twelve segments that each corresponded to an animal from the zodiac.

When I was a teenager, my mother told me: *Nothing good happens after midnight.* She was referring to my curfew, though instead of carousing I normally spent the hour of the ox in bed, praying I'd fall asleep, pressing the button on my projection clock to see how much time had passed, thoughts crashing.

She was right, though. The hour of the ox is the most night that night can be.

During the hour of the ox on October 27, 2018, I go into labor. During the hour of the ox on December 28, 2018, my father passes through the fabric.

Dying is similar to being born, my mother told me, during the many hours between those two events when we all live together, our family juggling the needs of two people whose weights are increasing and decreasing at rapid rates. *Both are a struggle.*

———

Before the hour of the ox there are other, brighter hours.

That summer and early autumn before my daughter is born and my father dies, Aaron and I nest. My father refers to the baby as Mabel during this time, since we don't share the name. We plan to move in with my parents for a few weeks after the birth.

I come up with a birth plan. Besides Aaron, I want my mother, the bulwark of my emotional support, next to me while I labor, and I want my father in the waiting room the entire time, ready to burst inside as soon as our daughter is born.

All my medical memories are inextricable from my father. When I broke my arm at age six, I crawled under a chair to avoid the nurse with the IV. My father lured me out from under the chair and onto his lap. He told this story later—his voice fluting up and down—emphasizing how much I cried, and how, while I was carrying on, hiding my face in his chest, the nurse stuck it into my vein without me realizing. I don't want it, I wailed, which is when he told me it had already happened.

He laughed when he finished this story, looked at me ruefully, as usual: *my silly daughter, whom I love.* The moral of that story, for him, was that my fear of things was always worse than the thing itself.

For me, the moral of the story was that if my father was there, it would be okay.

When I tell my father of my birth plan, he snorts. Why do I need to be in the waiting room the whole time? he asks. As a doctor, he's done his share of obstetrics rotations and knows how long labor can last. As a cancer patient, he has no desire to spend extra time in the hospital.

I'm just going to kick up my feet and watch TV at your apartment until Mabel's head pops out, he tells me.

In the end, my plans are moot. When I am thirty-eight weeks pregnant my father's PICC line becomes infected and he has to stay in the hospital until it clears.

My sisters and I text our mother a barrage of questions: *What is sepsis? Is he okay?* (*Okay,* now, is relative. *Okay* means he will not die immediately. He will never be the old kind of okay.)

On October 26, Aaron and I visit my father in the ICU, where the tightness of the nurses' faces and even the quality of the overhead lighting broadcasts a mood so unlike that of the ward that houses my prenatal appointments. There, we listen to the baby's heartbeat and say hurry, hurry. Here, we say: wait, wait. Stretching the rubber band of time.

All things considered, he is in remarkably high spirits, showing off his PICC line and giving me pointers for the birth. He tells me this is not the thing that is going to kill him.

I leave reassured, though by the time we arrive home forty-five minutes later I am not. I sit on the toilet, reading articles on sepsis-related death, until I notice a clear, sticky globule in my underwear.

I pick it up with a tissue and call for Aaron. Is this my mucus plug? I ask. Discharging the mucus plug is supposed to be one of the first signs of labor.

He grimaces. He doesn't know what it's supposed to look like.

Yes! That's the mucus plug, my dad texts back immediately, when I send him a photo. *You will go into labor soon!! Eat food!* Later I send him a photo of the pizza Aaron orders. *Good!!* he texts from his hospital bed.

I start to prepare myself for the fact that my mom won't be able to be with me, since she needs to stay with my dad. But she tells me to call her when I go into labor. I'll come, she says.

———

Some ancient Catholics believed supernatural activity was highest in the middle of the night due to the lack of organized prayer during that time. Others believed it was because Jesus was crucified at 3 P.M., marking that hour as his, and so the opposite time, 3 A.M., belonged to the devil. (A study once tested whether people sense "presences" most at three, due to the peak in the body's melatonin production. The results were insignificant.)[1]

In any case, the hour of the ox is the favorite time of yūrei, which literally translates into "faint [or "dim"] spirits." Yūrei generally refers to human souls that cannot cross peacefully into the afterlife, due to reasons like lack of proper funeral rites or overwhelming emotions like anger or envy, and is, in most ways, analogous to the concept of ghosts, though what kind of spirits belong in this category is debated.

A defining feature of yūrei—and one that distinguishes them from other yōkai—is that they maintain their previous human shape.[2] They retain their memories, personal histories, demeanors, and names. They primarily haunt people they know. There are numerous stories of someone meeting a friend in the street and having a perfectly normal conversation. Only later to find out that friend had died a week, a month, a year ago. What they'd encountered was a yūrei.

Shortly after midnight my contractions start, each about ten minutes apart. At this point they feel like a bad period cramp, pain I have known before.

In bed, I tell Aaron to go to sleep. I tell him at least one of us should get some rest, though as soon as he falls asleep not a minute later, with the lights on, I am irritated that he has done what I told him to do. Through the hour of the ox, I write in my journal. *When will she come. Hurry up. Hurry up.* When I moan, he half wakes, enough to squeeze my hand or rub my back, and falls back to sleep. For the next three hours he sleeps in intervals of six minutes then five then four.

Around three in the morning, I call the midwife.

You sound too chipper, she says. It's not time yet.

Finally at five in the morning—the hour of the rabbit, when light is not yet leaking through our windows—I call again. The hospital admits me; I am, apparently, no longer too chipper.

I am five centimeters dilated. Wow, the nurses say. Won't be long, they say. (But time has its own dilations, its own ideas.)

When my mother arrives forty minutes later, ping-ponging from one hospital to the next, I still have not yet decided on the epidural, my fear of needles temporarily overcoming the pain until the pain increases such that the calculus changes.

Before the epidural, I need an IV. My veins are secretive, hiding from this nurse the way they once hid from the nurse at my childhood physical until, defeated, she called in my father, who was not my doctor but knew the landscape of my arms. He pinched at my elbow. Here's a juicy one, he said, inserting the syringe in seamlessly, juicy like a nightcrawler.

Here in labor and delivery the nurse calls in a backup nurse, who jabs again and then requests the head anesthesiologist. I am blamed for moving around. I feel that everything is my fault, that I cannot make my body behave, as the nurse chides me with each poke. Even after they find a vein I cannot stop crying, only calming when they insert the epidural into my spine, a blessedly painless procedure.

Labor stalls. My body is a numb, joyless blank. It is the hour of the dragon and then the snake and then the horse, and I creep half centimeter by half centimeter. My mother calls my father, my sisters, updates the extended family. Aaron and I work on our *New York Times Thirsty Thursday* crosswords. We wait. I sleep. I am so hungry. The hour of the sheep, of the monkey.

In the evening the midwife finds meconium in my fluid, and suddenly four or five other nurses and doctors enter, and it's time to push.

I push and push and push. It is not working. We have to turn off the epidural, my midwife tells me grimly, you're not pushing hard enough. The next two hours I endure the worst pain I've ever felt, all bearing into my left hip. Her head is facing the wrong way, her head is stuck. The nurses and doctors crowd around the bed. I switch positions onto all fours, onto my side, the nurse jams his fist into my hip, this is the only moment of not-pain. They put a mirror in front of me. There is the crown of her head.

(The midwife does not tell us what she knows, which is that the umbilical cord is looped behind the baby's neck, like a ribbon folded in half. She does not tell us until the baby is out. For that I will always be grateful.)

After ninety minutes of pushing the midwife digs her hand inside me, unloops the cord, and there, at 8:18 P.M., the hour of the dog, is our daughter.

Her eyes closed, skin flushed, like a peach or an oni. Because of the meconium she is immediately handed to the doctors. I thought I would cry but I do not cry. My mother cries for me. The doctors do things in the corner that I cannot see and, finally, she is placed on my chest. She is impossibly small, this body of my body.

My first thought is of my father. We video-call Kristi, who is with him on hospital duty, and she holds the phone in front of him as he raises his glasses to squint at the screen. This is your granddaughter, I say, holding her up to the phone camera, she's named after you.

No calls, a nurse says, pushing the phone away, this is the skin-to-skin bonding time. You can do that later.

She does not know how little *later* there is.

————

Over the next couple days, I marvel at our daughter, who like me is born with jaundice, another tiny girl seeking the sun. Her little body, plump and thin-skinned as a gyoza dusted with rice flour right before being dropped in the pot. Her meager wail. Her lack of a Mongolian spot. Her whole self is a mystery ready to be known, or at least guessed.

While pregnant I'd worried that Aaron had never even held a baby before, yet here he is, every hour, fixing her swaddle when I've given up, holding her thigh steady when the nurse takes her blood.

Three days after her birth, my father is discharged from his hospital, and we are discharged from ours. Aaron and the baby and I emerge from the fluorescent lobby into the afternoon light. We drive out of the city alongside the maple and oak trees and their orange and crimson leaves, then onto the expressway. With the baby in our back seat, each thing I see is new.

When we pass my neighborhood's most significant detail—the strip mall graveyard where the ghosts of my childhood live—I tell her, this soft thing in the car seat that seems more like a wrinkle of fate than my child, that we are almost home. I use the word *home* to refer both to my parents' house in the suburbs and my own apartment in the city. In my mind, both spaces coexist.

How little we know when we pull into my parents' driveway, unload our suitcases, our Pack 'n Play, and our new child. We think we will stay for a few weeks. Instead we move into my old bedroom and do not leave for another two and a half months.

II

When my father holds his granddaughter for the first time, on that green couch, under that yellow light, his sings into her ear: *Chou chou, my little chou chou.* We nuzzle together, smelling of two different hospitals.

A few days later we take her on her very first walk. She is such a tiny thing in that stroller, her head a nub in a sea of blanket. My mother collects fallen leaves—yellow, orange, a brilliant red maple—to save in her baby book; her middle name means "autumn leaf." My parents walk in front of us holding hands. On this warm day, my father is wearing an enormous winter coat that envelops him in a sea of fleece.

This first month at home, this November, he is relatively healthy and mobile. He goes to work every day while Aaron and my mother and I stay home. I am tethered to the radius of the baby's bed and her bassinet and the chair by the window where I pump. This house is her, and my, entire world. The hour and days slide together with no demarcation except for her eating and sleeping schedule. What is night when we are up all night, what is day when we sleep in snatches in the brightest sunlight?

And the breastfeeding is so hard, her latch so painful, and her weight gain so minimal that I have to supplement with formula from the beginning. I wonder if it is worth it to continue nursing, since I can't take my Adderall until I stop. The stimulant can contaminate the milk supply. Sometimes I can forget my father is dying, I'm so busy trying to keep our daughter alive.

At first my sisters are in and out: a week here, a weekend there, then they move in and do not leave. Kristi gets a month of family and medical leave from her tech job; Cori can do much of her arts and nonprofit work remotely.

I start to refer to our home as the Death House. Which is a joke, as in the phrase: *Death House, no rules.* As in the answer to the question *Is this thing acceptable for us to do?* The answer, in those months, is always yes. You do what you must to survive.

Once upon a time, those of us in the Death House would have been marked as defiled. The gods hated pollution. This was known. A woman's blood, a child's birth, a person's death—anyone touched by such things had to stay ashore. The village could not risk letting a family member tainted by death or birth onto their vessels. None of us would have been allowed on a fishing boat.

Yet I think we are a circle unto ourselves not because the aura of death makes us profane, but because it makes us sacred.

My sisters watch *Brooklyn Nine-Nine* with our dad, their laughter carrying throughout the house. Kristi with her head tucked in the crook of his neck, all three covered by the same blanket, their feet poking out.

We take turns lulling the baby into slumber. We eat the food that people leave on our porch. I get irritated at my sisters for the tiniest things—not clearing the table as often as I think they should, not coming down to say hi to a guest. They get irritated at me for telling them what to do, for trying to control the situation, for falling into the old patterns. We tell stories.

We lie on the couch, heads leaning on shoulders. My father and I compare the vertical lines on our bellies: his old surgery scars bisect his entire torso, and my fading linea nigra, the dark mark on my stomach that appeared during pregnancy.

In my parents' house, it is just us and the television and the couch and his medicines and his jokes—always, until the end, his jokes—and my daughter, crying, and my father, crying, and all of us, crying and also laughing. Not my daughter—she doesn't know how to laugh yet. One day she will.

(I think of these as the bright hours. My memory is selective; time and recall infuse each moment with a rosy hue.)

After one month of paternity leave, Aaron returns to work, right as my father's work schedule becomes more erratic. He is tired. More and more he has to leave early from pain. He is frustrated by what he sees as his body's—his own—weakness.

One day I ask him, What do you think it's like after you die?

I think heaven is like New Zealand, he says. It's like I'm going on a vacation and you can't come with me yet. I'm going to stake out all the good places, all the best rinky-dink spots to eat.

My parents had visited a few years ago for their anniversary, sending us photos of them holding hands against vast undulating green slopes, a sea the mirror image of sky. I'd scanned the pictures for elves and dwarves but they hid outside the frame.

———

By Thanksgiving his death becomes legible in the crust forming on his lips and in his protruding clavicle. In my youth I dreamt of owning such a clavicle. Now, I see it for what it is: the final ossification.

My father and Kristi go to Hawai'i, his last fishing trip. He is in so much pain on the way home she is afraid they will have to go to the hospital.

When he returns, he has little appetite; everything causes pain. He is the incredible shrinking man.

If I eat a few hundred calories a day, he tells a friend, I have about three weeks left.

My mother spends her time cooking the few things he does want to eat. She boils his chicken and daikon soup and brews his doujiang from scratch, a multiday process that involves soaking, cooking, and grinding the soybeans, mixing them with water and peanuts.

Dust to dust, as the Bible would say. He now has the same amount of hair as my two-month-old daughter, only hers is as soft as a rabbit fur coat. His is like steel wool.

―――――

Yūrei—the spirits that appear during the hour of the ox—have the power to travel wherever they want.[3] The other kind of ghostly spirits, known as obake or bakemono, are generally fixed to a specific location. *Bakemono* means "changing things," while *obake* today is a more childish word, the one used to describe white-sheeted ghosts in my daughter's children's book. *Obake* is also the word my grandfather used as a child when he saw an eerie light across the slough.

Because obake cannot leave their locations, humans can avoid them by not visiting these places, Yanagita Kunio believed. What physical limitations they have, however, are expanded temporally. Unlike yūrei, obake can come out at any time, not only during the hour of the ox.

These definitions, of course, like yūrei, are not fixed. "In truth," writes Ikeda Yasaburō, a scholar, folklorist, and essayist, "the spirit realms are far too complicated for simple classification; any rule or distinction you make is immediately broken."[4]

―――――

Then my father stops being able to work. For him—a man who has loved his role as doctor, as caretaker—this is the beginning of the end.

By mid-December even a trip to the local Japanese grocer ends with my father coming through the door, clutching his stomach, doubled over. This is when the Death House becomes our boundary. Only Aaron still goes to work. The rest of us do not leave.

Instead, people begin to arrive. The California relatives fly in, all of his in-laws, who've loved him like he's their own. (Once, when introducing them to a guest, my mother's mother referred to my parents as *my son Charlie and his wife, Donna*.) In the videos of this visit, people are jammed together on the couches and on the floor. My father is in the middle of everyone, laughing and laughing.

My father's own family—my ama and aunt and uncle—come weekly. My ama still thinks he will be cured. His extended family plans to visit in January, but January never comes.

———

Every day his energy wanes, and more people want to see him than he can handle. My mother frets about turning them away, so I start to manage his inbox and social calendar.

We receive our guests in the living room, as if we live in the nineteenth-century Jane Austen novels my mother, Kristi, and I love.

Each visit is a portal into different moments of my father's life. One old friend brings a photo from college that shows my dad between two of his best friends. Both these men, who'd stood up in my parents' wedding, are dead now: one in his twenties in a bicycle accident, the other in his fifties from a sudden heart attack. And soon my father will be dead, and this photograph, as sepia tinted as my father's old aviator glasses, will be a reminder only of ghosts.

———

We joke that Dad is Frodo carrying the Ring that is his cancer. Mom is Sam, and Cori, Kristi, and I are the hobbits Merry and Pippin—both a help and a hindrance. When we try to watch the first movie one last time, we never reach the end. My father doubles over in pain, moaning *Papa, Papa*, a word I have never heard him use for his own father, who has been dead for over twenty years. He refuses more fentanyl.

If I take the drugs, I don't know what is going on, he says. I want to be awake for whatever I have left.

———

This happened, and then this happened, and then this happened. I cannot tell the story a different way. Cannot shape it. Is this the truth of death—that it cannot be crafted, no matter how many times we go over these lines?

———

We watch for signs that our father is dying, is dead. We have a little blue booklet from hospice that outlines the stages. Sometimes I am sure he is on death's door—his swollen legs propped on the couch, his mouth hanging open—and then there he is an hour later, in his red plaid Christmas pajamas, vacuuming the entryway. I take a video of this, captioned with something like *Classic Dad*, and post it to my Instagram stories. My mother tsks me when she sees it, saying it'll give people the wrong impression. She's right: messages pour in congratulating his health.

And yet a couple days later he can no longer go up and down the stairs. Our outer boundary shifts from the Death House to my parents' queen-size bed, where we all can, if we squish, lie down together. He leans into my mother, their heads turned into each other, and all three of us girls, and his granddaughter, curled up by their legs, trying not to kick each other. Our father strokes our hair like we are children again.

———

On December 23, 2018, a Sunday, my father wants snow crab and steak from Bob Chinn's, a surf and turf owned by a Chinese American man my grandfather knew back in the old Chicago days. My mother had suggested Bob Chinn's earlier that month, hoping it would tempt his appetite, and he declined. Today he wants it again.

He eats one bite of the New York strip steak, one bite of the snow crab. How is it? we ask; he has not wanted to eat anything for days. Delicious, he says.

After our mom leaves on an errand, he asks for what ends up being his actual last dying wish: for Kristi to steam clean the carpets in the upstairs living room.

She needs to learn, he says. He shows her what solution to use, where to find the spray bottles. It is, we understand, a ludicrous request, but he is dying, and cleaning has always been his self-soothing mechanism. We comply. He directs Cori and me to move the furniture into the hallway. Though I don't remember him ever moving them before, I default to following his directions, as I have my entire life.

He can barely walk. He leans against the couches and tries to push, despite our increasingly panicked exhortations: Dad, sit down. Dad, *sit down!*

He never moves those couches when cleaning, our mother says when she returns. He needs to lie down, she says. We pull him to the sofa, but he keeps popping up like a jack-in-the-box.

It reminds me of my grandmother, who lapsed into dementia so slowly that on several visits to California I took her increasingly bizarre instructions at face value. I did not understand that her mind was leaving us. What takes her years happens to my father in a day or two. In both situations I cannot recognize the crossing until they have already crossed.

———

That night, my father summons us all into the bedroom. We climb onto his bed, my mother and sisters and I. Aaron holds the baby in the chair nearby.

I love you, he tells us, his eyes barely open. He smiles down at us and takes turns kissing our foreheads. Thank you for taking care of me.

One of us makes a joke—*us* take care of *you*? *You* take care of *us*.

I cannot remember if I knew this was the last time. The events of the day, his energy, have tricked me, though the blue booklet clearly says there is usually one big bout of energy right before the end. This is supposed to be a portent, though when it is happening you think, Maybe tomorrow will be like this, and tomorrow's tomorrow.

On Monday morning, our father—the one we had known—is gone. He cannot speak. He sleeps most of the day. I cannot tell what or if he can see.

For the next five days, we are in limbo. These days are the worst of all: when I am hoping he will die.

———

We watch over his body as it stiffens and shrinks into nothing, like the last leaves in December. His gums recede. His bones protrude. There is nothing fetid about it, no blood, no vomit.

We no longer visit him all together. During this limbo time we enter individually, in the dark. I lie next to him in that bed, holding my daughter in my arms. Both of them asleep, though their sleep, I am certain, is different.

I try so many times to take a photo of them. I do not want to ask anyone to take it; the desire seems grotesque and self-serving. Still, I want to capture that memory. I hold her in my arms and try to position the camera but the room is dark and the photo blurry.

On Tuesday it is Christmas, and we open our presents in the bedroom in front of him. His eyes are wild as we show him our gifts, his jaw hanging open like a broken nutcracker. He makes squalling noises that sound first like the wind and then like a petulant toddler until my mother tells us we should go out into the living room; she thinks this is upsetting him.

My mother hands my sisters and me our final gift from our father—a wooden door hanger for each of us. On the front of mine, there is a photo of my dad and me in Toronto, and on the back, a note he wrote nine days earlier. His message, in his recognizable doctor scrawl, starts from the past (*I really enjoyed our dad and daughter time in Toronto . . .*) then sweeps into the future (*I will miss seeing how God continues to evolve you . . .*) and finally ends in the present, with his sign-off—*I Love You, dad,* next to a tiny heart scratched in ballpoint pen.

This love is the thing that remains both in the immediate present and in the historical present. It is the pin in the fold of the fabric of time.

On Wednesday night, after tiptoeing around all day as not to disturb him, we are sitting in the dimly lit living room when he appears in the doorframe of his pitch-black bedroom, lurching like Frankenstein's monster. It is the first time we've seen movement in days. Someone shrieks. I cannot remember if it is me or Kristi or if it is only in my head.

He comes out for a moment, those wild eyes, and then reenters his bedroom, propping his body with one arm against the wall. His limbs move erratically. *Zombie Dad,* he would have called himself, had he still been able to conjure the thought. (Whatever energy it is that animates this body, I fear it as much now as then.) My mother and Cori help his jerking body back into bed.

On Thursday, when my mother talks to him in bed before falling asleep, he reaches out and touches her hand. When she tells me this story the next morning, I am shocked. His body now seems like nothing more than a hub of frightening, autonomic responses, and yet he can comfort his wife of thirty-two years. My husband is in there, my mother says, and he still cares about me.

On Friday morning, our mother says we no longer have to be quiet. Dad, wherever he is, is no longer bothered by such mundane things as external stimuli.

In the middle of the night she appears at my bedroom door. There is a shadow against a crack of light. I am discombobulated. I think it happened, she says. She'd been awake, listening to his breathing grow louder and more ragged, and then it stopped—just like the little blue book had told us. She knocks on my sisters' doors, and we go to him.

It is nearly 3 A.M., the hour of the ox. My own daughter is asleep in her bassinet. We look at his body in the bed. It looks the same as before. My mother closes his mouth. I cry. I think, *thank God.*

———

In the morning my mother calls hospice. A nurse arrives at five or six in the morning to officially record the death.

This nurse, this stranger, appraises my father's body, decked in his Christmas plaid pajamas, laid out in my parents' bed. His hair, once black, then gray, all gone. His skin reptilian.

What time did he pass away? the man asks.

Around three, we say.

The man is surprised. Wow, he says, admiring my father. He looks back at me and says, You people don't get stiff as fast.

At these words, my father changes into a corpse before my very eyes.

What is there to say in response to the nurse? There is nothing. My father, apparently limber, is lying in front of us, dead, and I have no anger left inside me, only a very deep fatigue.

———

When our pastor comes later that morning, he suggests perhaps delaying the funeral due to church scheduling conflicts.

Another week! my sisters' and my eyes say to each other.

I think we need to have it soon, my mother tells him. My daughters need closure. They need to be able to go home.

We do not write an obituary. There are too many things to handle—primarily our emotions—and we do not know what to say.

After the funeral, we leave one by one—Cori back to Minneapolis, Kristi to her apartment in Bucktown. Aaron and I carry our things out of the house and into the car just as we'd carried them in nine weeks earlier, until only my mother remains in our home, our father's ashes resting on the mantel.

———

There are several ways a yūrei can move on. If they are only missing the proper funeral rites, they can convince a living person to perform them. If a grudge prevents them from crossing over, they can have a descendant avenge their death. Thus, the soul, now at peace, can cross over to the afterlife.

Those are the positive outcomes. In some cases, yūrei lose their essential humanity, and can no longer be considered yūrei. Then they are simply monsters.

Other yūrei are fated to stay yūrei forever. They will always belong to the hour of the ox. They will always beg to be remembered.

"Yūrei maintain their past existences while constantly attempting to insinuate themselves into the surface of the present," writes scholar and professor of literature and philosophy Yasunaga Toshinobu.[5] "This is not only a powerful rejection of the tendency among the living to forget the dead but also a desperate counter-strike against the living who would simply lay a beautiful veil over the past."

———

So: this is a ghost story.

A month after my father's death the temperature drops to -23 degrees, the coldest Chicago has seen in over thirty years. Ice collects along the windows of our apartment, whose heat we cannot control.

We call my mother, drive the forty minutes to my parents' house, where it is warmer. (*My parents' house—I mean, my mom's house*, I say the first year. *My mom's house—I mean, my parents' house*, I say the years following.)

It is our first time sleeping there since the funeral. Without the dead or the dying, it is no longer the Death House, though when I walk through the door the air is heavy with my father's absence. Or perhaps I mean his presence, for the home is nothing but the space that contained him: here, the spot on the couch where he completed his charts every night; there, where he recited part of "The Walrus and the Carpenter"; over there, where he danced with my mother.

That night I follow the patterns I know so well. Eat dinner with our mother, clear the table. Arrange the pumping station in my childhood bedroom. Assemble our daughter's Rock 'n Play next to my side of the bed. Wake intermittently to feed her.

In the hour of the ox someone calls my name. In the first hypnagogic moments I think it is again my mother, come to tell me that Dad is dead. How many times can a man die? I wonder. How many times can I learn he is dead? Is he stuck here somewhere, in his bed, unable to go down the stairs?

But it is only Aaron tapping my shoulder, telling me there is water streaming from the pipe under the bathroom sink. My father wasn't there to remind us to let this faucet drip overnight.

When I wake my mother, she is curled on her side of the bed even though the whole thing is now hers.

On the first floor, water streams from the ceiling down along the front door, directly below the upstairs sink. When we open the door the cold pushes in like a fist. Rivulets of water have formed into icicles zigging up and down the screen. Water pools on the welcome mat. Where the ceiling meets the doorframe the drywall sags and sweats.

What should we do? we ask. My father, the person who would know, does not deign to answer. Aaron googles *how to turn off pipes*. My mother finds the valve and turns it off.

In the hours before a plumber can come—*Sorry*, they say when I call, *there are pipes bursting all over Chicago*—the sky brightens but the air does not warm. The ceiling bubbles and drips and water seeps everywhere. We soak it up with towels and bedding from deep inside the linen closet, unearthing some I hadn't seen since my childhood: the giraffe towels my aunt had embroidered with my and my sisters' names, the queen-size duvet covers my parents had used throughout their marriage.

The water's all caught in the ceiling, the plumber says when he arrives. We need to release it. He looks at us. Is that okay?

Yes, we say. We do not know if this is the best option but know that we do not know any better, and here is someone with an authoritative voice. He cuts a slit into the middle of the sagging area. Water streams into the bucket below.

For nine weeks we lived in this house. One thousand five hundred and twelve hours, some of them bright and some of them dim and some in which we could not see. For nine weeks it was the boundary of our world, the place within which my father was alive, and then dying, and now is no longer.

"The feelings of the living toward the dead are what perpetuate yūrei culture," Komatsu Kazuhiko explains, "and what this portrays is actually the world of humans."[6]

After a few moments, the plumber takes our broom handle and pushes it up into the wet drywall, jabbing our ceiling again and again, and the water spills and spills, until at last he whips out a knife and cuts an enormous hole. Stand back, he says. And we do. We stand back and watch as all the water caught in the walls rushes out.

KETSU

In the end—

In the end, as always, what you think is simple is not simple.

Writer Imai Takajirō uses the word *musubi* to describe the final part of kishōtenketsu.[1] *Musubi* literally means "the tying up of loose ends." (*Musubi* is also your word for rice ball, when you fluff the steaming rice and pack it into a neat triangle before wrapping it in nori so it won't fall apart.)

But of ketsu, Yang Zhai says: "This verse should appropriately 'fade out,' and should end on a note of suggestiveness."[2]

A fade-out is not a jump cut.

There are always ends that will flap in the wind.

BAKU

The chimera who live deep in the forest
and feast on nightmares.

MOURNINGTIME

———

In the borderlands I call out for the baku, the swallower of nightmares. I don't think it can hear me all the way from Chicago, so three times I whisper, Baku-san, come eat my dream, baku-san, come eat my dream, baku-san, come eat my dream, as the old stories instruct.[1] In Japan, you see, the baku lives in the deepest forest, and here in Chicagoland most of our old-growth canopy is gone.

Though *gone* is really just a euphemism for dead.

stage one.

> *The first phase of the Ryūkyūan disposal method*
> *consists of the natural disappearance of the soft*
> *parts of the body, effected by exposure in the*
> *jungle, a cave, or a tomb.*[2]
>
> —ERIKA KANEKO

In parts of the Ryūkyū Kingdom, mourners used to wash the bones of their dead in a ceremony that took place over decades.[3] Some communities still carry out versions of this ritual of senkotsu today.

Here in the suburbs of Chicago, when two white men enter my parents' bedroom with a stretcher and a black body bag, they tell us to leave.

He's been dead only a few hours and already they're shooing us away. We retreat into the upstairs living room, peering through the door; they tell us to go downstairs. Multiple times they tell us this. At first, I think they want us out of the way of the stretcher. Later, I realize they do not want us watching at all.

We shuffle backward down the stairs. They are alone with my father, and it does not seem right.

I do not see my father go into the bag. I do not see them zip up the bag. Instead, we stand in the entryway and watch these men carry my father out the front door of our house.

The next time I see him he is in an urn on our mantelpiece. After that, I only see him at night.

———

In America, the funeral is often the dead's only event, organized quickly, when grief has not yet had time to settle into your body. A week after a loved one's death your body is still splintered with shock. We do not have the extended rituals other cultures have to commemorate their dead, after time has passed and your body has begun to understand.

At my father's funeral, held in our church sanctuary—still decorated for Christmas with red poinsettias and candles, along with all the arrangements sent by relatives, friends, and patients—my daughter sleeps in the crook of my elbow. In the aisle in front of the altar my mother and I stand in black, receiving all the visitors.

Your dad, they say. Your dad was the only one—

I bask in the glow of these stories while they are still vibrant, and practice the eulogy in my head. No one in my family, least of all me, thinks I'll get through the speech without breaking down. Behind the pulpit, I try channel my father's energy.

Hello. Most of you know me as Jami, Charlie's oldest daughter. To my father, I was rarely Jami. To him, I was J, or James, or beautiful, or chou chou da pigu—his favorite term of endearment—which in Mandarin means big, stinky butt.

During the speech tears gather but they do not fall. The grief has not yet been fully absorbed. In the speech I talk about my father's humor, his generosity, his faith.

I do not talk about how for so long I waited for the angels to bring me good news, for the stone to roll away, and how, on the third day after his death—the last day of the worst year—the tomb remained closed.

After the funeral, the guests stuff their faces with tuna maki and gaze at the baby with joy.

At least, someone says. The circle of life, another says. (My father had eight weeks with my daughter—four when he could carry her, four when he could not.)

Well, he got to be a grandfather, my doctor, who knew my father professionally, tells me. *It's not a bad way to go out.*

But for me there is no *at least.* In the weeks after, I will read many terrible self-help articles. Searching for something tangible. Do this and you will be cured. Do this and you will have the energy to lift your head, to face the day. The articles just say: *Grief is just love with nowhere to go.*

———

Later that year a family friend calls out *Charlie! Hey, Charlie!* across the church fellowship hall, and because it has only been ten months (how has it already been ten months?), I turn to look for my father. But he is dead, and not even in church can he be raised.

By *Charlie* the man means my baby daughter, crawling along the church carpet. She is not yet aware that this word means her, or that she is named after an agon she will never remember. Only this friend at church calls her by this nickname and for this I will forever be grateful. For speaking my father back into life week after week.

How is your new house coming along? the friend asks, after rubbing my daughter's head.

Aaron and I have recently purchased a home. I thought novelty might open the tight fist of grief. But once we move in, I discover my body is the same old body in a new house.

I shrug. Still a lot to do, I tell him.

Our basement is filled with unpacked boxes. The walls are white. The shelves are stacked on the floor. I am waiting for my father to come over and help me.

I see my father on his back, below a jury-rigged kitchen table he constructed out of an old door and Ikea cast-offs, a power drill in his right hand, his left searching for a screw. The whir of a drill, the crack of wood punctured and releasing, our shared laughter, dry shavings spinning to the floor.

When I try to hang a shelf, I use the wrong kind of anchor. I cannot pry it out with the pliers despite my best efforts. I look at the wall. I look at the hole. I abandon the task.

The problem is each of my days is tucked in another, larger day, like a series of Russian nesting dolls, and when each doll is cracked open my father is still dead.

———

stage two.

> *The second phase [of Ryūkyūan burial]*
> *necessitates human agency in disposing of any*
> *remains of the soft parts of the body, the so-called*
> *"bone-washing" ritual. The cleansed and purified*
> *bones are reassembled into a bone-jar and*
> *deposited in a cave, on a cliff ledge, or in a tomb.[4]*

That first December, we have one too many stockings. Advent is a season of preparation, but this year I am stuck.

I do not know what I am anticipating. My father is dead. I am thirty years old, and my father is dead. I seek movement but I cannot find any.

At a book launch, I read an essay that appears in a miscarriage and infant loss anthology.[5] I do not cry while telling the story of my miscarriage itself but when reading the part about my father I choke up so much I cannot breathe. For many long seconds I stand by the lectern and try to control the heaving that runs through my body. It seems obscene to be crying about my father's death at an event dedicated to a different kind of loss.

A year has passed, and I grieve the way I grieved yesterday, and tomorrow I know I will grieve the same. The only good thing about this time is I do not have to ask myself why I feel like this, the way I have during other times of despair. I know why I feel like this.

If yōkai are a process of naming, then having a name provides some relief.

Though only a little. My father is dead, and all my prayers are to my father. I have little to say to God.

The first Christmas, my mother and sisters and husband and daughter and I fly to California. My grandmother's dementia means we can't all stay at her house, so we ping-pong from place to place, feeling to feeling. In the afternoons, there is an undercurrent of tension; at night, each mouth is a lash or a stone.

On the anniversary of his death, my mother and sisters and I are all upset. We do not know how to grieve together. The family member who could smooth this over is no longer here.

We come together only right before we go to sleep, all of us crying, sitting on a queen bed in an Airbnb. Unable to agree on how to commemorate the day, someone suggests watching a TV show, our lowest common denominator.

Even then we cannot decide what to watch. Aaron is sitting on a chair and the baby is asleep and my mom and sisters and I are squished together like the four grandparents in *Willy Wonka*, or the way we were in my father's final days. We throw up our hands. We sit there. I am ready for the whole family to crack apart and fall off a cliff.

I have something we can watch, Kristi says finally.

She pulls up a forty-minute video of an art preservationist using a scalpel to chip yellow polyurethane coating off an old painting's surface so that he can restore what is underneath. The video is titled *Scraping, Scraping, Scraping or a Slow Descent into Madness*.[6] We watch him scrape forever. We stay in bed.

———

It is hard for me to talk about the bone-washing ceremony because it is so easy to exoticize, even for descendants and diaspora ourselves. To essentialize a specific ritual that was not universal. To take a ceremony and say *this is who they were, this is who I am*.

The story of this ritual was not passed down to me by my Okinawan great-grandparents. I learned it via research. From my family I had learned only about the soul ascending to heaven, the body remaining as husk. I believe there is strength in other kinds of sacraments.

In senkotsu, "rotted flesh is stripped away, bones cleansed, and the disorder and collapse of decay rectified. Their useless container discarded, the bones are carefully arranged—reconstructed—in a new vessel, strong and enduring," writes Christopher T. Nelson in *Dancing with the Dead: Memory, Performance, and Everyday Life in Postwar Okinawa*.[7] "Senkotsu transforms the decaying body, a thing of horror, to the subject of power encountered by human actors."

At least three years elapse in between the first stage of traditional Ryūkyūan disposal (when the flesh naturally disintegrates from the body) and when the community returns to collect the bones and wash them.[8]

Three years after his death, the grief that I thought would last forever has softened. During our yearly commemoration of him, I can cry without anguish. It does not hurt like it used to, and that seems like a new kind of loss. This is a cliché and it also feels true.

I can no longer remember how he smells.

I try to write about him but in the stories my father does not speak. He does not even appear—not really, not in any scenic way. I cannot describe his face or his smile or his laugh or the way his voice could— in an instant—bring my skittering heartbeat back to level. I worry that some will feel this as a lack. (It *is* a lack.)

I want your father to be more fleshed out, someone says after reading an early draft. To this I say: Me too. But I do not have flesh; I only have ghosts. In this story, the dead are only what I say they are. Does this make them less real?

———

stage three.

> *The final stage consists of the deceased losing his*
> *individual identity—on the physical plane by his*
> *bones being emptied onto a general ossuary and*
> *on the spiritual plane by joining the anonymous*
> *host of ancestral deities.*[9]

The third phase of Ryūkyūan disposal happens thirty-three years after death, when the mourners move the bones of the dead from an individual jar to a platform that holds all the bones of the ancestors.[10] Their spirit joins the collective.

Thirty-three years would be double my lifetime, one year more than the length of my parents' marriage. I wonder what memories will remain then, if he will still return to me.

———

Three years after my father's death, I begin to have terrible dreams. Thus I call for the baku, the swallower of nightmares.

The last animal created by the gods, the baku was formed from spare parts of the other creatures. Hence its elephant tusks, oxtail, bear hump, tiger feet, rhino eyes. Hence its desire—its need—to stay away from prying eyes. People love a chimera. (What I mean is people love finding a thing that beggars belief, and then killing it.)

The baku originates in China, where it is known as mo, or mengmo. There it scared away pain and disease, and the Tang emperors gifted their people its skin as protection.[11] When it came to Japan it adapted to its new environment. As the baku, it began to eat all the bad dreams.

In the moment before the moment when sunlight cracks across my face I'm always about to drown or die—or worse, die of embarrassment. Every morning I wake with my baby hairs matted around my crown, slick at the nape of my neck, water in my eyes.

What I mean is I'm desperate.

I google *baku talismans* online. The shipping is prohibitive. Instead I try to convince the baku itself to cross the sea for me. I make my backyard an accommodating home: I forget the weeds, let all the maple leaves accumulate, the branches grow wild, hoping that the baku will find a dark place to nestle and suck away my nightmares. The baku is the size of a tapir; the compost pile could hide a tapir.

After weeks of letting things go, all I've managed to attract is a yellow notice on the front door. It is the parks department telling us to cut our lawn within five days, or we'll have to pay a fine.

———

The baku, too, lives just outside the field of vision. I have never seen one in person, though I hear they have small heads and stumpy legs. Some have fur of gold like the sun and others silver, gray like moonlight. Their poop is a ten on the Mohs Hardness Scale—it can slice through a weapon—and their urine can disintegrate iron.[12]

I do not have bamboo or snakes or most of the other foods they like, though I do have metal tools. I scatter these across the backyard in what I hope is an appetizing way. Before I go to bed, I look out the window searching for a shadow in the garden, a hump in the compost. But it's not there, or if it is there, it's not listening when I call out, Baku-san, come eat my dream.

———

In this one, I'm looking out the window at a little girl standing on the apex of a shingled roof. A smokestack belches behind her. I'm counting to three the way I do when I'm trying to get my daughter to listen. On *two* the girl lifts one foot and shifts her weight sideways; she falls, her body horizontal through the air. I hear her splat on the ground. In this one, it is my job to tell everyone she is dead.

In this one I am standing on top of a recycling bin and a pack of dogs are nipping at my bare feet. I can feel their teeth pressing against my toes.

In this one and this one and this one, my teeth are falling out, leaving behind a trail of pearls, or bread crumbs.

When my husband wakes, he is restored; when I wake, I am weary, lashed with emotion that no one understands. No one wants to hear about your dreams. Dreaming vividly is a little like living a second emotional life, or watching a dramatic TV show every night—the characters and situations are not real, only the way you feel is.

My daughter does not dream yet. What a life, I think. Or perhaps the baku comes for her alone, knowing she is young and must be protected.

One night, when the moon is bright in the sky, I go to my window. Come find a dark place to nestle, I whisper. You will be safe here.

In this one I am onstage, and I have forgotten the choreography. In this one I have to return to my high school because they've lost my transcripts. In this one and this one and this one, I can't find the correct room number and the bell is about to ring. (I never have any dreams about college. I am always trapped in high school.)

The problem is each of my dreams is tucked inside another, larger dream, and when I wake, I am still inside. I cannot escape.

Why don't you stop taking your melatonin? a friend suggests. It got worse after you started taking it.

Then I won't sleep at all, I explain.

But it's something else, too.

No one wants to hear about dreams. So, let's say these are folktales. Let's say these are myths.

Once upon a time there is a girl who is lost on the edge of an island while the water rises around her. First, it's at her ankles. I could get home, she thinks, if only I had planned better. Then the sea is at her knees. She does not have a ship to take her away. Before she can drown, here comes her father with his planner and his airline miles. I've made a schedule already, he tells her. He points her to the airport and carries her away.

Once upon a time there is a girl in a car she cannot steer. The wheel is locked. But it's okay, because her father appears in the passenger seat, his voice low and soothing, and takes the wheel from her.

Once upon a time her father leans across the dryer in a damp basement she has never seen before, keeping her company as she hangs clothing across a line.

Once upon a time they stand in her parents' living room. He is turning on the light. When they hug, she is shocked to feel his body, the heft of it, the way this time her arms do not feel bone. You are so substantial, she tells him.

What is real: the way my heart thumps and my chest lurches and aches.

———

I never dream of Dad, my sisters tell me.

My mother has had a dream where a wave pulls her and my father under. They manage to hold on to each other, and emerge, their hands clasped. But this is rare; she too usually does not dream of him.

I think my subconscious just knows he's gone, she says.

My subconscious has not gotten the message. He reappears again and again, getting me out of scrapes, walking down the street with my mother. Once in a while he is invisible, making his presence known only within a thick cloud of dream logic.

Lacking a body, this is the way he returns to me.

Which is why I've been trying to lure the baku to America. I don't want all my dreams to go away. I only want it to eat the nightmares.

The nights pass, and the baku does not come no matter how many times I call. It knows what I cannot admit: That a pile of leaves cannot keep it safe. That no matter how big I grow my garden, or how wild I let it spread, it will never be enough to hide the baku from prying eyes.

Okay. Okay.

———

Once upon a time, four years after he dies, a mother and her three daughters wait at an airport gate for their father to arrive. Everything in the airport is red, like Lunar New Year, and they scan this color for a glimpse of their father. Finally, they spot him across the airport: his ticket has been printed wrong; the gate agent will not let him through. He is gesturing to the agent and waving at his family across the room. The girls ask each other if they should board without him.

He'll make the next one, their mother says. Get on. So they board the aircraft, sit in their assigned seats. They buckle their seat belts but are still fretting. He'll join us on the next plane, their mother says. So they wait.

I tell myself perhaps it is better that the baku never came. Calling the baku is not without risk; the baku can be overeager. It is so hungry and so alone. It can accidentally swallow into its sunken body all your good dreams and all your daydreams, your ambitions and desires, until there is nothing left.

The key to creating a successful character, I've been told, is to figure out what it wants. But sometimes after the baku comes, you do not want anything.

Lacking my father's body, lacking my father's bones, I wear his clothes. His red sweater, his gray plaid pajamas my uncle bought him when he was dying. He was so tiny at the end. They fit me well. In my prayers to God, I pass on messages. I say his names: Papu, Dad, Agon. I place an orange and a stick of incense in front of his photograph on my mantelpiece, though I know he preferred canned mandarin oranges to fresh.

Sometimes the hunter and the hunted share the same hunger. Sometimes the longing is the closest we can get.

I sleep, but my heart is awake.[13]

AMABIE

The mermaid-like yōkai whose image can prevent illness.

THE NAMING

———

I

For the longest time, when we said her name, our daughter wouldn't turn her head. I'd call out the name that was also my father's and instead of looking up, she'd keep doing whatever she was doing before.

C! C! C! I called into the abyss. I wanted to tick a box on a list.

Motherhood, I'd learned, was a journey whose guidebook was the CDC Developmental Milestones, a website and app with a list of all the things your child is supposed to accomplish by a certain age.

I found the video clips of babies modeling each task useful because I did not understand what actually counted as *pushes down on legs when feet are on a hard surface* or *shows fear in some situations*. Fear, the CDC told me, was a twelve-month milestone.[1]

Each appointment with her pediatrician:

Did you look at the milestones?

I looked at the milestones.

Is she meeting the milestones?

She is not meeting the milestones.

At which point we'd discuss what milestone she was missing, and the doctor would tell us to wait. By the next appointment, she usually caught up. But the doctor's *it's fine* did not have the gravitas of the CDC, which linked their milestones pages to lists of developmental delays. Such information—useful for early intervention—could be, for a certain type of parent, a boulder on your back. Your child is two weeks behind, your child is six weeks behind, are you doing enough, you aren't doing enough. In the parenting forums, mothers wondered whether their seven-month-olds had autism.

Those days there was always something to fret about. There was, for example, the recurrent problem of our daughter's constipation, which every month or so became the struggle around which our life circled. Many days I thought more about poop than I thought about my father.

And there was the problem of language: she understood all we said, and babbled to herself, but could not accurately copy our sounds. The only thing she could say was the noise a Japanese dog makes. *Wan-wan,* I'd say, seeing a dog in the street, and she'd repeat it, pointing with her tiny hand, so soft and little, *Wan-wan.*

The pediatrician said at fifteen months she'd give us a referral to speech therapy. The specialist said wait until eighteen months. Aaron's mother said not to worry—Aaron didn't talk much until he was three years old. The Internet said to panic.

Papa, Aaron told her. Can you say *Papa?*

—, she said.

Every day I woke up and my father was dead. Every morning this was news to me. The cold shock of it, like when, in the middle of the night, our daughter let out one piercing shriek.

All my journal entries began: *Dad*. All my prayers were secretly, blasphemously to him. When people at church spoke of the Heavenly Father, I saw his face. His tanned, speckled skin. His transition lenses.

Where are all your words? I asked our daughter one afternoon. Who has taken them?

She looked at me, silent. She rolled a ball underneath the couch for the thirty-eighth time. Those days, neither she nor my father nor God deigned to respond.

———

The question *When will she know her name?* was also, of course, *When will she know who she is?* For me, this felt especially fraught.

When I married Aaron, I kept my last name, wanting to maintain that visible part of my identity. For C, we carefully chose two middle names—the first was both Hebrew and Japanese, and the second was Lin. She had Aaron's last name.

At first, I thought her four names were a beautiful compromise. As she aged, I grew more uneasy. Her first and last names, the most salient parts, appeared completely white, in a society that hugely privileged whiteness. It began to feel like a loss of the part of her that was me, the part of her that was our history.

Aaron agreed when I asked if we could make Lin one of her last names, though we didn't take the plunge immediately. I was daunted by the bureaucratic process, and friends warned me about the dreaded double-barreled last name (*It will take so long to fill out forms!*). When I was feeling indulgent, I thought about bestowing Nakamura upon her as a fifth name.

Despite the complications, I could not stop thinking about how names are fundamental parts of our identities. They function as shorthand, as synecdoche, as a way to find each other. They cannot explain all our experiences, but they can gesture toward the things that bind us.

Yōkai narratives also reflect the process of naming. It is harder to tell a monster's story when you do not know what to call it, or what it is. In a way, a name bestows existence. Naming is a "concrete way to make sense of the worlds we contend with every day," Michael Dylan Foster writes. "A particular yōkai can be the metaphoric embodiment of a vague fear or mysterious phenomenon, like mists of anxiety merging into a sentient, tangible shape with a name and personality."[2]

When I was growing up, I knew about individual yōkai—the oni that Momotarō killed and the tanuki that could turn into a teapot—but I did not know the word *yōkai*. Without the name, I had no scaffold upon which to gather this information, and I did not know they belonged together in a category.

My children's book of Japanese folktales even translated all the names of the individual yōkai into their closest English variants. Oni was called "ogre"; in other picture books, "demon"; I must have known they were called "oni" because of my mother. By not naming these characters in the stories, their history was severed. To me the word *ogre* recalls trolls under bridges and other European tales. *Demon*, on the other hand, is immediately read within a Christian context. Neither captures the essence of the oni.

To name a thing is to see it more clearly. From Toriyama Sekien's first yōkai encyclopedia in 1776 to the book in your hands now, we try to filter through the chaos.

———

Eventually C began to respond to her name, though when her speech did not progress, I talked to the instructor of our weekly baby play class about early intervention. Overhearing, another mother approached me as I was packing up to go.

Your daughter isn't talking at all? she asked, pulling her mouth. I shook my head. Her own eighteen-month-old could say two hundred words and had learned to save herself from drowning, even with her winter coat on.

You should just talk to her all the time, the woman said, as I struggled to zip C into her own coat. C would not be able to save herself from drowning. In this coat, she could not even stand up if she fell over.

That's what we do, she continued. We're constantly talking to her and playing with her and singing. And we read all the time, too. Literally, we've read 3,500 books.

I was blinking back what I knew were irrational tears, iceberg tears.

I don't know if this is helpful or not, the woman said.

Thank you, I said. It's really helpful. I shoved my daughter's hat on her head; she was crying for her yogurt. At home I wept and texted my sisters and mother, who responded with sad face emoji and reassurances about my parenting. *We Hate Her*, Kristi said.

This was late 2019 and early 2020, the valley between our father's death and coronavirus, in which the things that could undo me were embarrassing and banal. My father died: yes, this was a monster everyone understood. My miscarriage, my subsequent fear for my daughter's life: yes, yes. Crises. Now, though, my mind circled obsessively over these things that I could not with a straight face call sorrows. They were irritants, fruit flies circling a drain.

Sometimes this felt like progress—to not always have grief in my brain like ice cream, churning and freezing—but only sometimes.

It was a gift, I supposed, that I had reached a point in my life where the major emotional swells of my week were limited to such interaction. Yet had my world become so small that this woman from the Friday 10 a.m. Baby Book Times class could take up so much of my brain matter?

———

One day I showed up to the same baby class and no one else was there. It was March 2020. I had read the news, I had seen the reports, and yet I didn't realize that I shouldn't have come; Illinois wouldn't issue their stay-at-home order for another few days. The class was held at the library that employed me, and they were still telling me to come to work.

Every other parent knew what I did not know: *do not bring your baby here.*

My fear is frequently misplaced.

II

During the pandemic, when all our communities were forced apart, we relied again on the amabie.

The amabie lives in the sea. It is safer, though not safe; she knows that no place on this planet is safe, not anymore. She enjoys the water while it lasts. The way her body glides.

She only gathers the gumption for shore every couple hundred years; the last time she appeared was in the spring of 1846.[3] Back then, officials in Kumamoto Prefecture in the south of Japan saw a light glowing in the water off their coast. Hmm, they thought. They rubbed their eyes—perhaps they'd drunk too much sake. The next night they saw the glowing thing again, and the next night, and the next, and finally they sent someone to investigate.

Luckily for us, we have documentation of what the official witnessed emerge from the sea: a creature covered in scales with long black hair, a bird's beak, and three finlike legs. She greeted the official. Hello, she said. Hello, said the village official. I'm the amabie, said the amabie. I live in the offing. She gestured to the water behind her with one of her fins.

Before the official could recover from his shock, she delivered two important messages. First, a prophecy: that the province would have a bountiful harvest for the next six years. The official rejoiced inside.

Also, added the amabie, with a swish of her three fins, when disease befalls your land—and it will—draw a picture of me, and show the picture to everyone. With this, she slipped back into the water.

Her message—and her image—was disseminated that year in the local kawaraban, a woodblock-printed newspaper. When the pandemic she prophesied arrived, her image was passed up and down the post road.

And then the disease passed, and everyone forgot about her for a century and a half, until she heard tell of another disease ravaging the lands.

Don't do it, the other sea creatures told her this time. They lolled around in the Dragon King's palace. The dolphins ate their shrimp cocktails, the shrimp their plankton cocktails. The world of men is no concern of ours, they said. They don't deserve you, or your gift.

It was true, of course. We didn't. But still something pulled her toward the surface.

While the rest of us were hiding away, quarantined in our communities, our homes, our bodies, quickly becoming more and more husk and less and less human, she again came ashore. She reminded people of her existence. Draw a picture of me, she said. Send a picture of me, she said.

What once took months down the post road now only took a second. During the pandemic, the image of the amabie was shared on five continents, and in tens of thousands of unique artworks and illustrations.[4] Her image became the face of Japan's Ministry of Health, Labour, and Welfare's coronavirus public safety campaign. Her image passed around again and again, across the globe. *Be safe. Be well.*

Scholars speculate her name is actually a misrendering of *amabiko*, another figure with a much lengthier and storied history.[5] But *amabie* is the name we now know this figure by, and it was the amabie's name and face that were shared around the world.

Her popularity partially depends on her cheerful image—even the original 1846 version was a merry one—in a dark time. But amabie is also an amulet. Amabie is a prayer or a hope made tangible.

I'd always believed in prayer but during the pandemic I began to believe in something you could hold in your hand, rub through your fingers. Those early months I kept a plastic bin in our yard where we could give and receive tiny offerings. Most of the people I exchanged with belonged to our Japanese American community, families with whom our friendship spanned generations. This person had extra rolls of toilet paper. That person delivered ten batches of cookie dough.

I gave away what I had from my garden: Japanese cucumbers, Ping Tung eggplants, our sun-dried tomatoes soaked in oil. I made my grandfather's recipe—Tom's Krunchy Tangy Takuan!—with all our daikon, and gave it away. In exchange: fifty-pound bags of flour, one pound of yeast, the messages going round and round—*What do you need? What can I give?* Each concrete item holding back the storm.

Our infrastructures were failing, but our people were not.

It reminded me of when my father was sick and meals would appear on our doorstep, months of dinners. Often you do not know what to say to the Death House. What is there, really, to say? Instead these bundles of food. *Be safe, be well.*

———

Around that time, my cousin Jay, who is white-presenting with a white last name, changed his middle name to Nakamura. My sisters and I were delighted with his transformation, another link that bound us. (Although names as links can be very general: we Lins share our last name with 8 percent of Taiwanese people.)

When I mentioned this, a friend told me her brother had recently dropped their father's white-presenting last name and switched to his mother's Japanese maiden name. He gave his daughter this last name as well. Something was in the air. These names as cultural preservation. A reversal, perhaps, of when my father's family emigrated to the U.S. in 1972, and my ama changed my father's name, Ching-Kuo, to Charles. (*Was it because the names sound similar?* I asked her once. *Yes,* she said. *And because he is a king.*)

I wondered if my grandmother had given my mother her own Okinawan last name as my mother's middle name—the way my mother named me—if it would have made a difference in our understanding of what it meant to be Uchinanchu.

Meanwhile, our daughter had five different cultural backgrounds and we were finding it a struggle to commemorate two or three. The information I found about multicultural families frequently touted the benefits of multiplicity, while overlooking the loss that comes from the reality that memories can only hold so much. Countless traditions are lost due to colonialism, forced erasure, migration, assimilation. On the individual level, they can be dropped due to lack of local community or available time.

In this way I knew names were both descriptive and prescriptive, and that no matter what my husband and I decided, our daughter ultimately held the decision in her hands. With more names than a law firm to choose from, in the future she would pick what she connected with and slough off the rest. She could choose to go by a nickname. She could hate all her names. As a teenager, I refused to tell my friends my middle name, presenting myself as Jami X. Lin on Facebook. My yonsei friends had middle names like Yukiko and Hanako—actual girls' names.

Only when I was older did I appreciate how Nakamura pulled me back to that grandfather who hopped on a cross-country bus a few years after the incarceration camps closed. Aiming for New York, he ran out of money in Chicago, where for a few weeks he lived in a bus station until he got word of a bar where Japanese Americans hung out.

There he ran into someone he knew from camp, whose mother let my grandfather live in her rooming house for six months rent free.

He landed his first photography job when he looked through the paper and saw an ad posted by a name he recognized—another person he knew from camp. That was how my grandfather eventually came to manage LaSalle Photo, a place that would employ many Japanese Americans over the decades.

Many people I've encountered through my research have had parents or grandparents who also were in Amache,[6] or who settled in Chicago after the war, and when I mention their names to my grandpa, he recognizes them. After I published an article referring to Amache, the niece of Grace Lee Boggs emailed me that her family had also been incarcerated there. I asked my grandfather, and he remembered them and which camp block they had lived in. The name, a way back to that memory, to that community.

———

In the third year of the pandemic, when our daughter was three, she started calling herself Aki-chan, a diminutive of her middle name. She did not understand that -chan was an ending used for all little children; she thought it was her proper name. It had transmogrified. I wonder how long each name will last. When she was born we thought we'd call her the nickname Charlie. But it wasn't her and it did not stick.

Still, my father's name remained in the air; she referred to him as if he were alive. I told her that he died, that he was sick. She did not understand. When we were planning for a vacation, she asked, Is Agon coming? When she found in her box of Moomin-shaped cookies one that was broken, she asked, Did Agon eat the head?

We did not call her by my father's name, but she called him by his.

———

The environmental management professor Takada Tomoki argues:

> *The fact that there are so many yōkai legends that have to do with natural disasters—in particular earthquakes, tsunami, drowning, floods, and shipwrecks—tells us that Japanese people were often facing these kinds of natural disasters, and were at the same time accumulating and passing down knowledge about those disasters and how to avoid them . . . this knowledge is contained in the yōkai legends.*[7]

Beyond entertainment, the yōkai tales give us guidance. I read the story of the amabie as one answer to the question *How do we protect ourselves?*: Turn to your community. To know its name so you can find your way back.

I think of the amabie image reproduced and exchanged digitally and physically all over the globe, all over social media. These tiny tangible acts.

My grandfather told me that when he was in Amache, his mother and other issei mothers followed the tradition of creating one thousand-stitch belts or pieces of cloth for their nisei sons, who were going off to fight in the 442, the Japanese American infantry unit. They would pass the fabric around to other people at camp, and each would make one stitch, or knot, with the goal of gathering one thousand stitches, so the person carrying it would have the strength of a thousand men. My great-grandmother made one for my great-uncle Joe. My grandfather and his siblings all sewed a single stitch. Afterward, they brought it around to the barracks, where all the members of each family in camp embroidered a stitch, each person adding their own version of hope, or prayer, or *be safe, be well,* into the slip of cloth.

RAT

—

The Rat, the first in the twelve-year zodiac cycle, symbolizes
diligence, thrift, and loyalty—and also opportunism.

THE YEAR OF THE RAT

———

PART ONE

LAND

My father was born a Rat. It would be better for this story if he died a Rat—a nice cyclical ending. His own father, for example, was born a Rabbit and died a Rabbit (in 1927 and 1999, seventy-two years apart), a fact I fixated on as a child. I understood little about my agon, who I saw perhaps once a week and could not talk to and whose body was usually affixed to a black leather couch.

Instead I focused on my agon's other material realities: the soft blanket he used to swing me, the sweetness of the condensed milk he fed me with strawberries. His blue toes, always in pain or falling off from diabetes.

His death occurring in the same zodiacal year as his birth pleased my ten-year-old's conception of pattern, a structure I could hold.

I know so much more of my father, yet in some ways he is still inscrutable to me.

I wonder how much of that inscrutability has to do with him growing up in a land that was not my own—there is so much water between Taiwan and the United States, and so much land between the coast and the Midwest—and how much of it was his personality, and how much of it was mine.

My father died at fifty-eight. He did not make it to the next Rat year.

In lieu of my father, I now have only fragments I must try to constellate.

———

In the summer of 2018, six months before he died, my family returned to Taiwan.

After my father was first diagnosed, he'd planned the trip for the following year—2019. My parents' therapist was more practical; her first husband had died of cancer.

Do it now, she told my parents. You don't know what will happen.

My father sighed. He didn't like therapy, was mildly grumpy about each appointment.

I don't know why I'm going, he told my mother, I don't have anything to talk about.

At which point my mother would remind him, again, that he was dying.

Vacuuming is my therapy, he said. Still, he attended out of love for my mother, who was the one who cried during their sessions. It's a three-hanky day, she'd update us in the family group chat afterward. Or sometimes: It's a ten-hanky day.

But when the therapist said, Go *now*, he listened. He rejiggered his frequent flyer miles, and all of us flew to Taipei in the Year of the Dog.

We'd visited Taiwan as a family only once before, when I was seven, for his brother's wedding. On that trip my father purchased a box of cookie sticks to entertain me on the rented bus we rode around Xiluo, my agon's birthplace. When I broke a cookie in half, I found a hair emerging from one end.

It's a rat hair, my dad said, holding up the strand.

I didn't understand he was joking—it was much too long to be a rat hair—and a few years later, when I was beginning to absorb the American cultural message of China (and by extension Taiwan) as a dirty place, this was what I remembered.

All of my other memories from that trip were also about food: the fluffy mantou dripping with sweetened condensed milk, the juicy sugarcane I chewed for an entire morning and that set me up for disappointment when, returning to Chicago, we bought a stalk from the grocery store, a limp haggard thing with tasteless juice.

Back home, I had no contact with Taiwanese America. My fourth-generation Japanese American identity, on the other hand, had been formed by decades of my family's generational ties to our church and the vibrant Chicago community that developed after incarceration and relocation. My best friends were yonsei, like me. And I grew up in the age of Pokémon, manga, teens flocking to our local Mitsuwa for Dance Dance Revolution battles. Japan was salient—and *cool*—to white Midwestern America.

But the only Taiwanese Americans I knew were my own family. And even that, identifying as Taiwanese, was a development; in elementary school, we said we were Chinese. This was true in a sense: by blood we were Han Chinese, though our ancestors moved to Taiwan in the early 1700s. Only when I reached junior high did we begin to call ourselves

Taiwanese, a political act spearheaded by my ama, who began to attend Taiwanese women's conventions with her sister and fly back to Taipei to vote for independence. I learned identity could shift—could, in some ways, be chosen.

But there were many layers to my confusion about *what we were*, most of them stemming from my ama's multiplicity of languages. I heard her speak Japanese, the language of her childhood under colonization, to her sister and to students she tutored in her home. When extended family came into town, she spoke Mandarin, the language she learned by force when Chiang Kai-shek and the Kuomintang took over Taiwan after the war. And though she told me she learned Taiwanese Hokkien when she married my agon, I never heard her speak it. With me, and with my father and his siblings, she spoke only English.

For all these languages, all these stories, much was lost in translation. I am still sifting, sifting, sifting, trying to make sense of what is between the lines.

———

The Rat is the first year in the Chinese zodiac cycle. The cycle comprises the twelve animals that participated in the Great Race. Long ago, the Jade Emperor desired twelve guards, and so he sent an emissary to the earth to send a message to the twelve animals. The animals must all race toward the Heavenly Gates, and the quicker they were, the higher rank they would receive from the Jade Emperor.

In the end, the Rat finished first. He accomplished this by leaping onto the ear of the strong, swift Ox, who carried him throughout the entire race, until the Rat leapt forward at the end, beating the Ox by a hair. The Jade Emperor rewarded the Rat by ranking him first in the race and first in the twelve-year cycle of the zodiac.

Cleverness is the Rat nature, Ama always told me. *Charlie is a genius,* she said. *Your daddy is genius.* She never mentioned any of the idioms connecting the Rat to dishonesty and cheating.

When I thought of rats, though, I imagined their wriggling tails and their role as vectors of disease. Of all the animals in the cycle, they seemed most of the earth.

———

Emperor Wen of Zhou said: "If you dream about a rat, you could have many enemies. If you dream of catching a rat, you will suffer treacherous plots planned by others. If you dream of a cat catching a rat, you will be blessed because your enemies will be weakened from killing each other."[1]

But my father was the most honorable man I knew. It is hard for me to critique him, to render the flaws that I hear make a character seem more human. He may remain flat.

Perhaps it is hard for me to critique him because it was hard for him, until the end, to show the world the soft parts of himself until the pain and illness, by necessity, broke that barrier. He swallowed so many things so that his children would not have to.

———

The stories I heard of my father's early adolescence in America— of him taking his neighbor's jeans out of the trash to wear, his parents struggling in their restaurant jobs, the other kids calling him *sorry Charlie*—made me ache for him, and I did not want to ask for more details. Lacking context, I formed a false conclusion from these anecdotes. I assumed my father fit neatly into the struggling immigrant narrative; I did not understand that our family back in Taiwan was wealthy, and how this wealth had shaped everything else.

To be fair, my ama told me stories about how her father had been an important doctor and local politician in Taiwan, but she always used the same voice as when she told me things like *Your daddy was the best swimmer in all of Taipei* and *I learned swords because my junior high teacher says it is better to be killed than raped* and *You and your cousins were all geniuses.*

I loved her stories and I got lost in them. When the KMT came, was it Second Uncle who had to dig his own grave, and Third Uncle who fled into the woods, or vice versa? She was my only connection to Taiwan, and all my knowledge was filtered through her. I knew little of the country's history. I did not know about the White Terror or the KMT or how any of it fit together. I did not know what was history and what was myth and what stories had been inflated by love and time and pride.

I did not understand, until I was an adult, how the Rat finished first because he leapt onto the Ox. The Ox carried him. I am sure the Rat never speaks of the Ox.

I am trying to focus the lens more, though it is still shaky, and my hand tired. That my parents were some of the most generous and kind people I knew, all their time and love and friendship passed on to others without expecting anything in return—this was true. And yet it is money that makes so much of what I do, and my family does, possible. My mental health would be much worse now if I lived under financial stress; without my expensive treatment in high school, I might not have gotten stable at all.

The way that money is so often the ox we ride until, near the end, we slip off and cross the line.

God will provide, say all the Christians I know. It is easy to say this when we lack nothing essential. (How to tell a story without demanding, from your readers, absolution?)

I looked in the Bible to see if there were any rat tales included but the only verses that mention rats in the Old Testament say not to eat them, and in the New Testament: now you can eat them.

———

As we traveled all around Taroko Gorge, a mountainous national park, on that last family trip in 2018, I felt that I had to swallow everything, for this was all the Taiwan I would ever get. My father was the link, my father was the mouth; if I did not soak it up now it would later be gone.

He sat in the front seat of our rental and conversed with our driver, translated for all of us in the back. My father's Mandarin was rusty— he only ever used it when visiting Taiwan every handful of years. He still could converse, though occasionally had to pause and ask a clarifying question.

I thought of the difference it would make next time we came, to not have an intermediary, to be able to say and understand nothing.

In a spa in the middle of the forest we soaked in a variety of baths, each one with a different element or temperature to soothe your body. We floated while my mother swam laps, her long arms longer in the water. In the pool, as water showered down from above, I asked him: *What was it like when—How did you feel when*—as if you could soak up thirty years' lack of knowledge in a week.

Later, when we had dinner with all his cousins, they all laughed and clapped each other on the back and complimented my father on his Mandarin and told stories of the boy I never knew. The time he fell down the stairs. The time their grandfather, who was himself a doctor, took the whole family out for a duck dinner because my five-year-old father had memorized the English alphabet.

After the meal my father came home and collapsed like a starfish on the hotel bed. Talking in Mandarin is exhausting, he said. Constantly translating back and forth in your head. My mother brought him a bag of crisps to rejuvenate him.

He tucked it in the crook of his neck and lay there. After a few minutes, I noticed and asked what he was doing.

I have a chip on my shoulder, he said.

He could be impatient, especially when we were lollygagging before a flight, but was never-endingly patient when waiting for someone to get the joke.

In Taroko Gorge I stood against the railing and tried to measure how high up we were.

I took picture after picture. It was impossible to capture the vastness of the gorge, or its depth.

———

What did you first think when your parents said you were moving to America? I asked my father on one of our walks.

I thought, What will happen to all my books? he said. He explained that my agon had encouraged his love of reading by developing quite a library for him, featuring characters like the Monkey King. I never knew this—I'd never thought of my agon as someone who would encourage books. My father's first worry, when he learned about their impending emigration, was about the loss of all those stories that did, indeed, end up having to stay in Taiwan.

Through extensive research, my translator Jenna Tang uncovered much about my family that I did not know. She discovered an old family plaque inscribed with the qualities the Lin family patriarchs hoped to instill in their descendants: mutual respect, harmony, self-reflection, frugality, moderate consumption so that the family wealth could grow, love and care for family.

As generations pass, information once gleaned through intimate relationships is replaced with research. This archival information does not portray my agon's quiet smile, or my ama's vehemence, or the way my mother claps her hands while laughing, or the way my father sighed with relief at the end of a day. An emotional disconnect.

I know there is a difference between reading about a place and going to a place and descending from a place and being from a place. When we walked in Taiwan, unlike in Japan, it felt like a land I loved but did not recognize. I did not know what to say without flattening or essentializing. The lands are separated by all that water between.

PART TWO
SKY

When people ask me about my sign in the Western zodiac—that system of stars and months—I know only enough to say Aries. I know nothing about the Ram; ask me about the Rat.

———

Once, Jenna told me, there were two mice who were great friends.[2] A snake ate one of them, and went back to his little cave to digest. The mouse's friend, bent on vengeance, went to where the snake was sleeping and bit his tail and held on tightly. The snake wriggled and wriggled, eventually shedding his skin and shaking off the mouse.

The next day, the mouse returned and bit on the snake's tail again. The snake tried to capture it, but the tiny mouse kept eluding its grasp. This happened for several days, and the snake grew more and more agitated. Finally, the snake threw up the mouse he'd eaten. It was not sitting well in his stomach.

Though the second mouse by this time was far away from the snake, he smelled his friend's scent. He returned to the cave and found the friend's body, soaked in the snake's digestive juices, and was filled with sorrow. He brought the friend's body back home.

What is the moral of this story? I asked Jenna. I did not get it. When she told me the story I kept getting confused as to which of the two mice we were talking about. When the snake spat up the mouse, I thought it was alive again: that the second mouse had rescued it and brought it back to life.

No, Jenna said, it's still dead. But the second mouse got to bring its body home. The moral of the story is to be a loyal friend and to have friends that will always take care of you.

Without her as an intermediary I would have read it and not understood. I had the text but not the context.

To have people that will always take care of you, even beyond this life.

———

I remember. I remember. I remember.

What is it I remember—each little story, each little memory, each little star?

The shape of a country whose status as a nation is debated and constantly under threat?

The shape of a man?

———

In the early twentieth century, the International Astronomical Union divided the sky into eighty-eight official constellations, based on identifications by the Greeks and earlier ancient civilizations.[3] There is Cetus, the whale. Corvus, the crow. Vulpecula, the fox. Hydra, the sea serpent.[4] There are so many animals, all the Western zodiac signs, but there is no rat. It is not marked in the sky. (I thought the word zodiac had some relationship to the stars, but it means "a little circle of animals.")

There are also groupings of stars called asterisms: "patterns or shapes of stars that are not related to the known constellations, but nonetheless are widely recognised by laypeople."[5] Stars from different constellations can join to be an asterism. Different asterisms can form a larger constellation. The Big Dipper—as a child, one of the only constellations I knew—is actually an asterism, part of the constellation Ursa Major.[6] (The word is related to *asterisk*, "little star," my second-favorite punctuation mark. To add, to clarify, to hedge a bet.)

In other words, asterisms are like constellations, only less well-defined.

The Chinese divided their sky differently from the Greeks. Their constellations have different names. There are the Three Enclosures (Purple Forbidden, Supreme Palace, Heavenly Market) and the Twenty-Eight Mansions. There is the Azure Dragon of the East, and the Vermilion Bird of the South.[7] There are the doomed lovers Ox-Boy and Weaver Girl, forever separated by the Milky Way, whose once-yearly meeting—facilitated by a bridge of magpies—is remembered in one of China's four great folktales.[8] They are celebrated in China, Taiwan, Korea, and Vietnam on the seventh day of the seventh month of the lunar year, and on July 7 in Japan.

And at the fundamentalist high school I attended, my philosophy teacher told me the Gospel was visible in the night sky. What the Greeks called Gemini—the Twins—were really the biblical Jacob and Esau.

Even the official constellations are, as the IAU itself says, "a matter of perspective. They are simply our Earth-based interpretation of two-dimensional star patterns on the sky made up of stars of many differing brightnesses and distances from Earth."[9]

Trying to form a pattern, a picture, out of what looks like a handful of sprinkles thrown on the ground. Trying to make this a person, trying to form an identity. When they conflict, when they conflate. When the space between one country and the next, between the past and the present, between the reality on this plane and the reality on the next—seems as vast as the heavens.

———

We were never a star-gazing family. The closest we came to watching the sky was when we flew through it. I was never interested in stars, though after college I developed an interest in the black holes that develop after one dies.

In graduate school I wrote a long diagram-laden essay about the black hole information paradox and the event horizon. I favored Wikipedia's definition of the latter term: "a boundary beyond which events cannot affect an observer."[10] I understood this as the demarcation between that which the black hole will swallow and that which the black hole will not swallow.

(*Black holes*, my mother says, *give me the heebie-jeebies.*)

Did I only care—did I only pay attention—when something was dying? When we were approaching the outer limits of the boundary? In Taiwan that summer I asked my father so many questions, and before that, so few.

While editing this essay, when I am struggling to form connections, I ask myself again and again, *But where is the Rat, I have lost the Rat.*

While editing this essay, when I am struggling to find the heart of my father, black holes appear once again in front-page news. Astronomers have finally captured a photo of our galaxy's own black hole: "a trapdoor in space-time through which the equivalent of four million suns have been dispatched to eternity, leaving behind only their gravity and violently bent space-time," writes a *New York Times* columnist.[11] I send Cori the link. *VIOLENTLY BENT SPACE-TIME!!* I text, all in caps.

(Our father loved science and science fiction. Though Kristi is the only one who adores *Star Trek* in the way he did, he passed down to all of us our love for sci-fi, our wonder for what is beyond.)

After numerous articles about the black hole photograph, Yvette Cendes, radio astronomer at the Center for Astrophysics, pushes back on a popular misconception, clarifying that they do not suck in everything around them.[12] They *alter*. This is called, she says, spaghettification. Not swallowing, not always.

I tell so many stories without fully understanding what lies beneath the surface. This is the limit of my present knowledge, the way I connect the dots at this single moment. I worry that filling in the gaps with research seems false. There is a nagging feeling that I should only tell the stories received orally from my ancestors instead of those I find in books. I think this is a common diaspora anguish, for those of us fractured from places, narratives. And yet without research, without looking beyond, we would be hamstrung only by what our ancestors shared. We can treasure what they gifted us, while also acknowledging that in certain cases they did not tell us enough, or that what they said was wrong, or was right for that time and place but needs to change now.

———

On that last trip to Taiwan, day by day we relished life, and day by day it relished us until by the end of the trip my family was exhausted. At the National Palace Museum, we fought the crowds to take pictures of the Meat-Shaped Stone and the Jadeite Cabbage and followed my father, who held his black collapsible umbrella high over his head in imitation of the tour guides' flags. At Chia Te the bakery was hot and crowded with people purchasing pineapple cakes and I knelt on the floor, my sun hat pulled low, while I waited for my parents to buy omiyage.

This was not particular to this Last Hurrah but a feature of our vacations in general: one day too long, one activity too much, go too hard, get too cranky, tired, needing a room with a door that shut.

This is, I thought, the end.

But for my father the trip held one last pleasure: the flight home. His body was deteriorating, and he used his miles to spring for a business-class ticket for himself.

Right before he died, one of the guys he mentored—the friendship between our two families spans four generations—sent a letter that I read out loud to my father, who was by then barely conscious.

The letter ended: *Enjoy first class. We'll hold down coach.*

———

After I am grown with my own baby, when we live in a suburb with less light pollution, I lie in our hammock on the nights when the Internet tells me to watch for meteor showers. I wait and wait but we live fifteen minutes from O'Hare, and everything I think might be a meteor is an airplane.

Years after his death, I still think of my father when I fly. When I was little, he told me that if a spaceship could break the speed of light, it would move faster than the earth, making us travel backward in time. On every flight, I think, *maybe if we go fast enough—*

The air: a place between countries, between states of being. Everyone on the plane knows from what location you've departed, but only you know if you are leaving home, or returning.

ONI

The mallet-wielding, tiger-skin-loincloth-wearing, fearsome
yōkai whose name is often translated as "demon" or "ogre."

IN THE WHIRLPOOLS

———

Mukashi mukashi—long, long ago—when you are a tiny girl in a big blanket, your ama calls out your name. *Jenko!* Jenko is her nickname for you, a rhyme of your father's nickname Genko. You are at your ama and agon's house, ten minutes away from your own, and they are babysitting you, this tiny crying girl swung in a blanket. The harder you cry, the more vigorously your grandparents swing.

This is their one proven trick. But today you are inconsolable. You miss your parents, who, you are certain, will never return.

Jenko, Ama says. You know the little boy who comes from a peach? *Big peach?*

She cannot use her arms to show you how big, since she has the blanket corners held tight in her fist, but the way her voice warbles when she says the word helps you imagine it.

You stop sobbing long enough to think of all the little boys you know. Some of them come from church, and others come from school, and the one you are most afraid of—the one who skulks around with a

BB gun and will, in later life, die of an overdose—comes from the neighborhood, but you do not know of any boys who come from fruit.

You know this story? Momotarō? she asks.

You nuzzle your face into the fuzz of the velour blanket and shake your head no. Your agon holds the other corners; your grandparents' muscles are taut for people in their sixties. They swing so high you are perpendicular to the ground. *Swoosh. Swoosh.*

Well, she says. Momotarō. Long long ago there's an old woman and an old man. They both want children, but they have no babies and so they are very sad. This is in Japan. One day the old woman is washing her clothes and she sees a peach—a *big* peach—coming down the river. Oh oh! she says. It looks so tasty. So she brings it home for the old man to eat.

When the old man cuts it open with a big knife, they find a little boy inside.

My name is Momotarō, says the little boy. The kami gave me to you as a gift. They know you wanted children.

Momotarō was so, so strong. Like you. Girls are just as strong as boys. Girls are strong like tigers. Momotarō was strong, and he *never* cried. When he got bigger, he decided to go fight the oni. Oni were very bad. They were big and they had horns and they liked to eat people. They came to Momotarō's village and took its food and all its treasure and ate its people. But Momotarō was not a crybaby, so he went to fight the oni.

For his trip, his mommy gave him some kibidango to eat. You know kibidango? No? Hmm. Kibidango is kind of like mochi. Yes, I know you love mochi. Now is not mochi time. If you listen, later is mochi.

On his trip, he met a dog, a monkey, and a bird. Momotarō shared his kibidango with the animals because Momotarō is very kind. The animals helped him sail to the island of oni. They had a big battle, the oni fought and fought, and Momotarō and the animals killed them!

Then he took all the treasures back home and the village was surprised and happy.

Now I teach you the song in Japanese, your ama concludes.

This is the way the story of Momotarō comes to you, the only folktale you will learn orally, from your elders, instead of through a book.

As Ama begins to sing Momotarō's song in Japanese—*Momotarō-san, Momotarō-san, o-koshi ni tsuketa kibidango*—the melody and the swinging fuse until they form one single beat, lulling you and lulling you.

You are nothing, a tiny girl in a tiny boat on a long river, rocking back and forth on the waves, listening to the song.

———

When you look out the side of your boat you see the thing bobbing on the river, like Ama said—that enormous peach. You can smell its fragrance even from here. You want to get closer but you don't know how to swim yet. You see the little old woman waiting on the shore with her arms spread wide. She gathers the peach in her arms and goes inside her little cottage. A moment later, a tall man emerges in samurai garb, wielding a sword.

Who is that? you wonder.

That's Momotarō, says a voice from above.

Grandpa? you ask. *You're* here now?

Yes, he says. And Grandma, too. Ama and Agon were tired of swinging. They needed help. Now we're all here.

Why is Momotarō grown? What about the peach? Where is the baby?

Time moves differently in these waters, says the voice of Grandma. Things happen quickly, she says. You have to be quick on your feet, or you'll be left behind. This is the classic Momotarō, the one whose story was told starting from the fifteenth century.[1]

You watch Momotarō climb into his boat. It's bigger than yours. His bird is brown. His dog is brown. His monkey is brown. As they sail away into the distance, it becomes one big brown smudge.

He's not as cute as when he was a baby, you say.

No, Ama says. He is not. But look, across the water—the oni! You see the oni?

You do see the oni, in the distance—enormous creatures, some blue, some red, some green. Even from here you can see their fangs, the horns emerging from the top of their heads, their tiger-skin loincloth, the heavy iron clubs in their hands. You hide your face behind your hands the way you do at Disney World when you run into the Beast or the Seven Dwarves or any of the characters you like in stories but not in person. You cry a little and lie down in your boat.

Maybe this was not a great idea, says Grandma.

You sail on down the river for you don't know how long. You ask for your parents. They'll be home soon, your grandparents say together. You cry harder, because time is funny here, and you don't know how long *soon* means.

Look, says Ama. There's Momotarō again.

You sit up. You are still in the river. Everything else looks different. On the bank of the river there is a man. The boat is different this time, too—bigger, more modern. The man is also bigger, with different hair, and he is carrying a white flag with a red dot in the middle. He looks angrier this time.

What happened to the other Momotarō? you ask.

Time passed, Ama says. This is the new and improved Meiji-era Momotarō.

Meiji era?

From 1862 to 1912, says Grandpa, who in all versions of time remembers all the pertinent facts and figures. It started when Emperor Meiji defeated the shogunate and took back the imperial power. Japan prided itself on having an unbroken line of emperors descending from the sun kami, Amaterasu. They said their power was from the gods.[*] Did you know that?

No, you say. I'm only a toddler.

Well, says Grandpa, if you don't know now, you will later. Watch out. The water is getting choppy. Lie back.

You lie back.

So then Japan sought to become a big world power, he says. They industrialized and opened their doors to the West. They sanctioned certain gods and beliefs and exterminated many others.

Pay attention now, Grandma says, seeing you start to nod off. This is where my family comes in. In 1879, Japan annexed Okinawa and the other Ryūkyū Islands and abolished the Ryūkyūan royal government.

In 1895, they colonized Taiwan, where *our* family lived, says Agon. This is more than you've ever heard him say.

And our people left, continues Grandpa. In 1898, my father—your great-grandfather—left Hiroshima for California. A year after that, Japan took the land of the indigenous Ainu people and banned them from using their language, their name, or their traditional ceremonies.[2]

---

[*] When Japan lost World War II, General MacArthur forced Emperor Hirohito to make a proclamation that he was, in fact, just a man.

Then *my* father left Okinawa for O'ahu in 1905, says Grandma. Her voice is much stronger than later, when she develops dementia, though that little lick of laughter at the end of her sentences when she talks to her grandchildren always stays the same. And, she says, five years later Japan colonized Korea, and so many other nations.

I don't understand, you say. I want a story again. Go back to Momotarō.

There he is, says Ama. See him, on that island far, far away? He's different now because the Ministry of Education printed his story in their textbooks for children like you.[3]

You squint. You can barely see the red dot of the Japanese flag waving in the sun. It's so far away.

In the new book version, says Grandpa, the oni were evil and worthy of conquering because they disobeyed the "benevolent rule of imperial Japan."[4] Momotarō was not only a hero, but a hero of imperial Japan. The textbook version presents a clear delineation between the heroes and the monsters, an easy way to teach children the difference between *us* and *them*. A clear *other* promotes a sense of nationalism.

I don't get it, you say. You begin to weary.

Grandpa tries a different tack. See how much bigger the oni are now? See how much more gruesome?

You peek between your fingers. Even at this distance, you can hear the oni's roar.

Who do you like better, Grandma asks, the oni or Momotarō?

Momotarō, you say. It is easy to identify with Momotarō. He is, after all, a very small, very cute boy in the beginning. You look at his pink cheeks and think *awww*. This is why his face is always chosen to adorn the cover of English-language Japanese folktale collections.

You hear an even louder roar that sounds like something dying. You cover your ears.

Down the river, Agon says, and the river speeds up suddenly, taking you swirling with it, spinning in a whirlpool that churns so quickly you pass out from dizziness.

Dad? you say, when you wake. Mom? Your shirt is slick with spit-up. You look down. Your arms are bigger, your legs. Your boat is rocking unsteadily.

No, the voice of Ama says breezily. Still us!

You look toward the voice. Now the sky is filled with planes and smoke. You can see, from far away, a newfangled kind of ship coming toward you—a ship so large it barely fits in the river.

We're in World War Two now, Grandma explains. During *that* time—

Wait, you say. Grandpa tells the history lessons. Where's Grandpa?

A pause.

He'll be here later, Grandma says.

Oh, you say. This does not make you feel better. *Soon, later*—the adults keep using these words, but they are all relative.

Anyway, Grandma says. Look, here's Momotarō again, back with his oni.

As the ship passes by, you see that Momotarō is even bigger than last time. He is wearing some sort of military outfit. His animal friends have grown huge in size, and are decked out in the insignia of countries: Germany, Italy.

During this time, Grandma says, the story of Momotarō developed as a nationalist symbol. Japan cracked down on all types of magazines to control the narrative. They backed one magazine, *Manga*, which published this comic that showed the oni in a different way. Oh, look, there he is now. Look, look!

You look, and on the far shore you see what looks like a regular oni, but as your boat gets closer, you see that the oni is a white man wearing a mix of traditional samurai and Western clothing. He has a gun, and he's shooting at Japanese villagers.

It's FDR, Ama says. Oh, Franklin Delano Roosevelt.

Is that real? you ask.

No, she says. It's a political cartoon. The caption says, *I'm the oni, I'm the oni.*[5]

Is Roosevelt the oni? you ask.

Well, says Grandma, according to this book by Noriko T. Reider—

Where's Grandpa? you ask again. Grandpa is the one who always tells me these kinds of historical facts.

According to this book, Grandma says more forcefully, at the same time as this propaganda was spreading quickly and deliberately, "the Japanese army was acting like oni in various Asian countries, their acts exemplified in the atrocities of Nanking in 1937, and the Bataan Death March in 1942."[6] There. Does that answer your question?

Where are my parents? Where is Grandpa?

Your parents will be back soon, Grandma says, and your grandfather— your grandfather—

You wait. Little wavelets knock the side of your boat.

He's in camp, Ama says. World War Two, remember?

Oh. You hadn't put two and two together.

Is that why the boat is so rocky? you ask.

Yes, Agon says after a moment. It's off-kilter.

Oh. I miss him.

I miss him, too, says Grandma. Even though out there in that time-life, he is a teenager and I haven't met him yet. But here, look, let me read from the writing he did at Amache, and it'll be like he's here.

It's not the same. Your voice takes on a petulant tone.

I know, says Grandma. But it's the best we've got for now. Look, here's a part when he's just arrived at camp and sees his barracks. I'll even try to read in your grandfather's voice: *I was wondering how will they ever put all of us in a small place that small. . . . What surprised me most was why did the soldiers have to stand guard with guns . . . and to tell you the truth the way some people stared at us it chilled me a bit.*[7]

Grandma?

Yes?

I don't like this.

I know.

Can we speed up? You look around. The current has come to a standstill.

The river of time does what it wants, Agon says. Our only luck is finding a whirlpool and we're not even moving.

I could read you something else from Grandpa's writing, says Grandma.

O-kay, you say. Only if it's not about guns or staring?

I could read to you from the memoir he's writing in *your* time-life.

Okay, look. Here's a whole page about how the toilets worked at camp. You know Grandpa, he loves to know the engineering of things. He writes a lot about gardening. Or here's a part about dancing.

I love dancing, you say. Tell me about dancing.

Apparently, there are a lot of dances for the young nisei at Amache, Grandma says. They still need to make a life. Look, Grandpa says: *The mess hall floor was concrete and with a non-skid surface it was difficult to dance. Someone from the kitchen got a bag of cornstarch and sprinkled it on the floor.*[8]

Dancing on cornstarch? You wrinkle your nose.

It's an incarceration camp! Grandma says. They have to make do! *The first event was more like a dance practice where boys would dance with boys and girls with girls until our confidence grew to a point where we gathered up enough courage to ask a girl for the first dance. With sweaty hands and clumsy feet, we were able to survive the first round.*

Maybe he's dancing right this very second, Grandma says. Her voice has a dreamy quality.

Would you have danced with Grandpa, if you were at camp? you ask.

I would, she says. Grandpa was a real looker.

So were you, Grandma. That's what Mom always says.

But not like him. Grandma laughs.

It's true. You've seen photos of Grandpa back in those days, during camp and after, when he was a navy photographer, and later, when he managed LaSalle Photo. To you, he looks like a movie star. But to them he was an enemy alien, and they were in charge, not you, so that is that. Your boat is rocky.

Finally, the current begins to pick up.

There's no more Momotarōs down the river, Ama says.

But there's oni, she adds. Let's keep going.

Agon steers you into another whirlpool. This time, you're prepared, and when you emerge you manage not to puke.

Hi Jami, says a voice.

Oh, Grandpa! Thank God.

Took a little detour there, says Grandpa. I'm back now.

But are my parents back yet? This seems like the longest movie date ever.

Look, there's your daddy on the shore, says Ama.

I mean my real parents, in my real time, you say.

Soon, Ama says. There we are, coming to America in 1972.

When you shield your eyes with your hand, you see a family of five getting off an airplane at O'Hare. The children are thin and their haircuts bowl-like and Ama is dressed as a very fancy lady.

I didn't know we could see *us* in the river of time, you say. I thought we only saw stories. Not living people.

Silence.

What? you say.

The river does what it will, says Agon. Look, there's your dad meeting your mother. Look at his goofy glasses.

Slow down, you say, but the current is moving too quickly. You almost lose your balance. You see your parents in glimpses, from behind the trees that map the shoreline: there they are eating pie on a date, there they are on their wedding day in 1986, there they are with a small yellowish lump in a fabric Easter egg—

It's me! you say. There you are as an infant, and there you are as a toddler, and there you are, getting rocked by Ama and Agon in their Palatine living room.

Is that us *right now?* you ask.

Depends on what you mean by *now*, Grandpa says.

Depends on what you mean by *right*, Ama says.

Depends on what you mean by *us*, Agon says.

Can we find another whirlpool? Grandma says.

What? you say. You want to watch this. You have waited your whole life to grow older, and now you finally can.

You shouldn't see your future, says Grandma.

Why? you ask. Could I change it?

The river of time is the river of time, says Agon. It does what it will.

Your boat swerves this way and that. Agon tries to steer.

Bring us to the next oni, Grandma says. Let's get out of here.

Which oni? says Agon, steering to the right. In the future there are so many oni stories. The narrative keeps fracturing. Which year should we go to? Do you want the oni stories of 2001? The oni stories of 2008? The oni stories of 2016? 2018?

Not 2018, Ama and Grandma and Grandpa say in unison.

What happens in 2018? Agon asks. I don't know, I'm dead by then.

You're dead? you ask.

Now look what you did, Grandma says.

You're dead? you repeat. Are you dead, too, Grandpa? Who else is dead?

Never mind, says Grandma. Let's skip past 2018. 2020?

Here's a whirlpool, Agon says, and in you go.

———

When you look out the boat there is a barge coming down the river. On the barge is one stage, and on the stage are two platforms, and on the platforms are two men, one white and one Asian. In front of the stage are hundreds of seats but they are all empty. You cannot hear what they are saying because they are shouting over each other.

Where are all the people? you ask.

Everyone's sick, Grandma says.

Everyone? you ask.

In 2020, there is a global pandemic, she says.

You are trying to wrap your head around what this means, but the men are so loud you can barely hear yourself think.

Steer closer so we can hear their words, Ama says.

Do I have to? Agon says. They hurt my ears.

Only for a second, Ama says.

The boat steers toward the enormous barge. You're close enough to the men now to hear their words. For some reason, their faces are even fuzzier up close.

DURING WORLD WAR TWO! the Asian man shouts.

This again? you ask. I thought we were in 2020.

What do you mean, this again? Grandpa asks. Like I'm writing in my memoirs—*The topping on the cake came when Japan attacked Pearl Harbor. It still remains in my mind as though it occurred yesterday!* I remember exactly what my barracks looked like in camp. I remember all the names of the people who were in Merced* with me: the Kitagawas, Oshitas, Komatsubaras, Fuchigamis, Fukumitsus, Nabetas, Tagawas, Hasegawas . . .

There were so many things I didn't talk about before, says Grandpa. After camp, so many things got muddled and lost.

DURING WORLD WAR TWO! the Asian man shouts again, sensing that he's been interrupted.

THE JAPANESE AMERICANS! VOLUNTEERED! FOR MILITARY DUTY AT THE HIGHEST POSSIBLE LEVEL! TO DEMONSTRATE THAT THEY WERE AMERICANS![9]

We *were* Americans, Grandpa says. Now my mother and father, they couldn't be American citizens because the American laws wouldn't let them be naturalized, but they'd been living here for decades—

Ahem, says the man. NOW—

When are we again? you ask. I've lost track.

2020, says Grandma. A global pandemic. Racism is high against Asian Americans because the virus started in China.

NOW! says the man. MANY IN THE ASIAN AMERICAN COMMUNITY! ARE STEPPING UP! TRYING TO DEMONSTRATE THAT WE CAN BE PART OF THE SOLUTION!

What solution? you ask.

---

* Merced Assembly Center, where my grandfather and his family were held temporarily before Amache was completed.

WE ASIAN AMERICANS NEED TO EMBRACE AND SHOW
OUR AMERICAN-NESS! the man shouts.

As he shouts, his face gets blurry. It loses and regains focus.

Why is he glitching like that? you ask.

He's not only one man, says Agon. He's a lot of men.

WE NEED TO STEP UP! HELP OUR NEIGHBORS!

WE SHOULD SHOW WITHOUT A SHADOW OF A
DOUBT THAT WE ARE AMERICANS WHO WILL DO
OUR PART FOR OUR COUNTRY IN THIS TIME OF NEED!

WE ARE NOT THE VIRUS! BUT WE CAN BE PART OF
THE CURE!

Can I steer away from this shouty man? Agon asks.

Yes, yes, your three other grandparents say.

Where is the oni here? you ask. I thought you said there was an oni.
Is *he* the oni? You look at the shouty man, still shouting, but growing
quieter as he recedes in the distance. Wait. Am *I* the oni?

You look at your footie pajamas. They are not red, white, and blue.

No, Grandma says.

Is that man very important? Is he the president?

No, Grandpa says. He is not, in the scheme of things, very important.
He is just a symbol.

Of what? you ask.

The idea of proof saving us, he says. They tried to tell us this during
the war, too, and after. That we should prove ourselves. Many of us
believed that myth. It still lingers.

If I don't do all those things that man says, you ask, am I part of the virus?

No, say your elders all together. We are not part of the virus.

We float along the river. It is slow now, and the sun is setting, a blazing yellow-orange sky. You scan the shoreline for signs of yourself or your family, but these days, it seems, everyone is inside.

Is this the legacy of incarceration? you ask. The way that man spoke about it?

No, says Grandpa. There are so many communities that use it to fight for solidarity. We don't always understand the terms, we grandparents. We don't always agree. It can be hard to talk about across the generations. But we are telling our stories in different ways, too. I'm writing my memoir, called *A Time in History*.

So, you're still alive?

As far as I know.

What about after you die? Where do all the stories go then?

What do you mean?

You sit up in the boat and try to figure out how to explain. All those stories live in your bodies, but what happens after you and my parents all die?

Don't worry too much, Ama says. We live for an extremely long time.

Not me, says Agon.

Well. Not you, says Ama. You weren't a big storyteller, anyway.

I had stories, says Agon. But my grandchildren and I didn't speak the same language. But look, he says to you, you can understand me now, right?

Yes, you say. And so when the rest of you die, I'm old enough to not worry about it?

Silence.

You throw your hands in the air. How will I remember everything from the whirlpools? you ask. I'm already starting to forget what I've seen.

You look backward. You have traveled so far. You cannot see World War II, much less the original Momotarō. The stories have been shaped by many different mouths.

One more whirlpool, says Agon.

When you emerge this time, you can barely see the river. Everything around you is shrouded in mist.

These are your last oni, says the voice of Agon.

Well, the first oni, says Ama.

Right, Agon agrees.

Where? you ask. You can see nothing, not even your hand in front of your face. Everything is light gray. Your pajamas are damp.

All around, says Grandma. In the earliest stories, oni were *invisible* spirits whose purpose was to cause physical harm and sickness.

To keep such epidemic deities from entering their capital, Kyoto citizens would conduct special rituals to give the spirits a physical form,[10] says Grandpa. They wanted to turn the invisible visible, and thus lessen their strength.

Oni are terrifying, but less terrifying, apparently, than spirits without faces, says Agon. People think that if you know what your enemy looks like, you can find them and defeat them.

When are we? you ask.

The beginning, your grandparents say. We made it back here again.

You don't understand.

Will you tell me who it is? you ask. What the enemy looks like? So I can warn my children, and their children?

Silence.

The thing is that the oni are dependent on the story, says Grandma. The story transforms with the teller. The stories one time-life needs are not the stories the next time-life needs. For we tellers—we can be wrong.

Even you?

Even us.

Even me?

Even you.

You sit with this for a moment.

How do I know which are the important parts to remember? you ask. The river is so long!

That's the thing about the river, says Agon. Every point in the river affects everything downstream.

The important parts are often the parts that keep coming back, says Grandpa.

You know what the important parts are when you go into the whirlpool and you come out again in the same place, says Agon.

Time, says Ama, like haunting, is a repetition. Something passes across her face.

Listen, says Grandma. The truth is we don't really know. We are not your real grandparents, out there in your time-life. We are only ideas. We can only speculate.

You are a child, and you do not like this answer. You look forward, into the offing, to where the gosei and the rokusei and the nanasei—could there ever be such a thing as nanasei?—live.

What is the point of the river if it provides no answers? you ask.

That is not the function of the river, Agon says. If you look to the river to give you something it cannot provide, you will always be disappointed.

What is the point of it, then? you ask. You can feel your lower lip trembling. These sound like riddles.

To row you to the questions, says Agon. And to the next question, and the next.

Oh, you say. Where are my parents? Your grandparents have been saying *soon* for centuries.

Here, they say. Here.

Your parents lift you out of your blanket. You are damp from mist and slick with spit-up and tears. Your father carries you outside to the car, your mother thanks your agon and ama. Here, your parents: solid, full of smells, ready to take you home.

———

In this lifetime, your grandfather seems to know everything that could possibly be known. He has a memory like a steel trap. You only have a vague idea what a steel trap is, but your grandfather is the kind of man who knows the exact mechanics of how such a thing works. When the battery of his flip phone died, he rigged up a D cell battery. You saw the wires and marveled even if you didn't understand at all.

When you were a child, your grandfather wrote and illustrated a book for you based on the time he saw an obake, a ghost yōkai, as a kid. This happened in Yuba City, California, before the family was incarcerated. The book features Yoyo, his fictional alter ego. Now, when your daughter sends him a card in the mail, he sends her back a drawing. *Hi C! My name is Yoyo! Will you be my friend?*

When you get the card, you get that scrunchy feeling behind your eyes. Not tears exactly, but almost. Like the pressure that builds before a sneeze.

It's the same feeling you get when your grandmother with dementia waves to your daughter and laughs, despite not knowing who she is, while being latched onto a big hook that moves her from her wheelchair to the couch.

It's the same feeling you get when you bring your daughter to your ama's house every other week, and your ama sets out the little pink chair and the plates of mochi and tea and when she mixes ink with water and holds your daughter's hand in hers, teaching her the strokes of different characters. *Mountain. River. Man. Mountain. River. Man.*

In the season of spring, your ama looks more alive, as bright as the peach-hued irises outside of her house, though her arms are still only as wide as your wrists.

Remember them, she tells your daughter. If you remember one, next time I give you prize. The prize is a box of Pocky, or puffy chocolate snacks in the shape of Pokémon.

It's the feeling you get when your ama turns to water her plants and your daughter runs up behind and clutches her legs.

Your daughter will never remember your father, but she will remember her great-grandparents. Time is funny like that.

A few years before he died, your father started stashing airline miles. There's going to be many funerals soon, he said, you'll need to go to

California a lot. He listed off the ages of your grandparents, their siblings. Instead, he was the first to die, and those other dates still loom in the distance.

Your grandfather, who loved your father like his own son, is ninety-four. Your ama, who talks to your father every morning, is ninety. Your grandma is eighty-eight.

Your daughter is three. You pray they will slow down, these ships passing in the night.

———

In this lifetime, there are so many things you and your grandparents never talked about—tales that disappeared. In this lifetime, there are some topics you don't talk about because your opinions are too different. In this lifetime, you learn the larger history of your nations, your family, only as an adult. You can only imagine what it would have been like to learn as a child.

In this lifetime, it is your sister Cori who tells you the stories about the river of time, your sister who does the work of tracing the intergenerational trauma that bends and winds through your family's history. It is Cori who designs oni handkerchiefs for the Chicago-based Nikkei social justice collective. She and Kristi wear them at their demonstrations. In this lifetime, you embrace the symbol, though your daughter is a little afraid of the image you display in the dining room.

In this lifetime, there are still so many rivulets to be traced back to the river's mouth. It is Cori who asks, How can *we* be good ancestors?

———

In an essay about the writer, director, and playwright Terayama Shūji, critic Steven C. Ridgely writes: "If myth and everyday life are not inseparable, then liberation cannot be found by . . . containing folklore through museumification; it can only be found by taking possession of the narratives by rewriting and remixing them."[11]

In other words: If the ghosts in the stories are the ghosts in our dreams are the ghosts in our closets, we cannot put them behind a placard, inside a glass box. The farther we get from the origin, the looser the narrative becomes. But perhaps what it loses in precision, it gains in expansiveness. The story is a different story. The story is the same story.

So maybe this isn't a story about ghosts, but a story about telling a story about ghosts. About how to remember while moving forward. You are drawn to these myths because they change. Unlike static texts, folklore, legends, and oral histories are living tales that transmogrify with each subsequent retelling. We understand and understand again based on contemporary lenses and critical frameworks.

The texts and the contexts mutually transform. The stories stay alive only as long as we keep telling them.

So when your grandfather—your only remaining direct male ancestor—comes to visit, you dig up the little picture book he once made for you, a story based on his own childhood. It included an aside about him being afraid of the slough near his home in California because the obake—another term for ghosts—resided there. How did you know the obake lived there, Grandpa? you ask him. What stories did you hear about them?

Oh, he says, laughing, I don't remember.

In lieu of his memories, you let your imagination wander, assemble snapshots from movies you've seen, fragments from books you've read, trying to create a scene to fill the gap where all the lost stories live.

————

Your grandfather has been working on his memoir for at least six years. In the beginning, he painstakingly typed it on his Windows 7, until his computer crashed, and he lost thirty single-spaced pages. He started again. He lost the document again, and he started again. When he tells you this, you set up a shared Google Doc for him. Occasionally, you download it to your computer just in case.

He did not tell these stories to your mother when she was growing up. He brought her and her sisters to Amache, but he never told her what it was like. He did not pass down the traditions from his own mother. Your mother had to do her own searching. She studied in Japan in college.

Only decades later, after your grandfather started attending the Amache reunions, did he start talking about what happened to his family. Your generation got to hear the stories your mother's did not.

After your father dies, your mother takes you and your sisters back to Amache, to that barren desert in Colorado. You are not prepared. You are wearing zoris, and your feet get all scratched. You carry your daughter on your hip as you see the guard tower that remains, read the signs, look inside the tiny barrack. (*I was wondering how will they ever put all of us in a small place that small.*)

It seems—all of it—impossible. This is the lineage of yonsei. You are here, but you were never here. You can only speculate.

TSUKUMOGAMI

Tools and other household items that have lived for ninety-nine or one hundred years and have gained a spirit.

AUTOMYTHOLOGIES II
(REPRISE)

—

Once, when I was twenty-five or twenty-six, I came home to an empty
house. It was five in the evening, twilight outside; I assumed my
mother was running errands and my father not yet back from work.
Upstairs I flopped on the couch and turned on the TV, as I had all
throughout my adolescence.

When the garage door whirred, I wasn't sure which parent it was until
I heard the slam of the door between the garage and the house, and
the clink of the keys onto our wooden chest, both movements heavy
enough to distinguish my father.

For some reason I did not call out to him. It still felt like my own home
then, though I no longer lived there, and had not done so for many
years. Instead, I dallied, getting a drink of water, going to the bathroom.
I did not say anything until he opened the door to the living room. Hi,
Papu, I said. He startled, eyes wide, emitting a guttural *hnngh* out of
his mouth and nostrils. A face of panic.

J, he said. You scared me. He leaned with one hand against the couch and the other on his heart.

At dinner he told my mother about how he'd forgotten I was coming over for dinner, about how he'd heard the noises upstairs.

Did you think I was a burglar? I asked him.

No, he said. I thought you were a ghost.

Really? my mother asked, surprised. We had never heard him speak of such things.

He only nodded, repeating: I thought you were a ghost.

———

When I was a teenager, I thought if anyone were to turn into a ghost it would be me. But now it is my father.

I cannot conjure him in this story the way I wish, though when he was dying, I kept careful records with the kind of details that might help render him human, our family real. It's nearly four years later and I still cannot crack those notebooks' spines.

The longer I let my journals sit unread, the likelier it is that they will transform into a kyōrinrin, the yōkai of the archive. The kyōrinrin is a birdlike dragon made of paper, with sutra scrolls for eyes and unfurled scrolls for a torso and arms.[1] It is the spirit of knowledge. More specifically, it is the spirit of knowledge that has been abandoned. The body of the kyōrinrin is formed from books and scriptures no longer read by their owners.

The kyōrinrin is a type of tsukumogami, tools or household objects that have acquired a spirit. Generally speaking, when these everyday items live to be ninety-nine or one hundred years old, they develop the ability to transform into animate beings. Objects discarded before that age are not granted this ability, and after becoming yōkai by other means, turn vengeful toward the previous owners who got rid of them.[2]

A century of accumulated wisdom gives the kyōrinrin much strength. It wields that strength against its foolish owner, who have allowed such wisdom to become dusty on a shelf.

The kyōrinrin first appeared in Toriyama Sekien's yōkai encyclopedia of 1784, which focused exclusively on tsukumogami, and was the follow-up to his original *Gazu Hyakki Yagyō (The Illustrated Night Parade of One Hundred Demons)* encyclopedia of a decade prior. That text was the first visual encyclopedic book of yōkai, though Toriyama was greatly influenced by the Hyakki Yagyō Emaki, the illustrated picture scrolls of the Night Parade that had been around since the tenth century.[3] Many of the yōkai depicted in these ancient scrolls are tools—umbrellas, teapots, and fans. In the scrolls, they appear joyful and mischievous, reflecting the way they cavorted down the streets of Kyoto during the Night Parade itself.

My father has undergone a reverse transformation: from animate being to household object. In my home he sits on the mantelpiece in a blue marble urn small as the palm of my daughter's hand, next to a candle and a roll of incense and a small mandarin orange. In my home he is pinned on the mirror above my desk, immortalized in a professional headshot from his first year of practice. He's wearing his aviator glasses and a suit—he looks like a goober; he never wore a suit. He watches me every morning when I take my medications.

In Cori's apartment in Edgewater, he lives on the windowsill that serves as a butsudan, a Japanese home altar, kept in the only place in her apartment that receives any light. In Kristi's apartment in Bucktown, he lives in the wooden door hanger she keeps on her dresser.

In my parents' home he lives in an urn my mother made in her pottery class. It does not look like an urn—all sharp angles, ridged textures pulled through the clay. It fits alongside her other handcrafted objects, all the pots and containers she's made to hold other things.

Any object in my parents' house could easily grow old enough to develop a spirit: they are kept, repaired, remade, so many times over. Every room is filled with the things he cleaned and mended. My mother's grief lives in that house they shared almost their entire married life, which always has space for us. My mother lives there daily. She has an app that tells her exactly how long it's been since my father died.

We keep him close, and when we travel to the places he loved, we sprinkle him. When I am dead, I, too, want to be cremated like my father and his father before him and his fathers before him. Among all the traditions lost, this one we have held.

Here in America, we privilege moving past grief. We are uncomfortable acknowledging the dead among us. Lacking communal, temporal traditions like Obon or Ghost Month or Day of the Dead, we seek individual ways to fill that gap. In my suburb there is a brewpub that moonlights as a business that turns your loved ones' ashes into diamonds. I get it. Without ritual, you need a tangible reminder.

So the archive, for me, has always been something I can hold when I can't trust my own memory—how little I can trust it!—or when memory provides nothing at all. I cannot be certain of the future but with the archive I can be reasonably certain of the past.

———

Two years after my father's death, on his birthday, we gathered with his siblings and mother on a video call. It was January 2021, the peak of the pandemic, though there were so many peaks they could no longer be counted. On Zoom we all lacked a body. We shared the stories we knew of him. All of us spirits.

Charlie, can you hear us, my ama said, at the end of the ceremony, speaking into the screen. It was a statement, not a question. No one responded.

I think so, she concluded.

I think so, too. I do not know so. But I know so little and, each day, I think I know less.

————

As I was wrapping up this book, I asked my mother to verify the dates of my father's hospitalizations.

Let me check, she said, I think I have his diary.

His diary? I asked. Something stirred. I had forgotten he'd kept one.

It won't be very exciting, she said. It's going to be, like, *Didn't get to go home today again. Didn't fart. I pulled out my own NG tube.*

A couple weeks later, she dug it up for me. It was a black and white composition notebook, like the first journal I kept when I was nine. On the inner cover, he'd glued a picture of his first tumor with the date of its removal: October 6, 2003.

I drove home with this precious cargo in the passenger's seat, unsure when I'd ever be ready to read it. But then my daughter fell asleep in the car, and I was stuck in the driveway with no other reading material. Okay, I thought, and I opened the cover.

From my mother's earlier description, I thought his diary would be a banal fact-finding mission. It was not. When I turned the page, first the shock of his cramped, indecipherable handwriting, interspersed with the looping cursive of my mother, to whom he'd dictate when too exhausted. (*Thank God for Donna*, he writes. *I'm such a mental case every time I'm sick.*)

The diary begins in the summer of 2003, before his first tumor. *I am going to start a diary to record what God is showing me and doing in my life*, he writes. One of the first entries is an exhaustive chart of all the fish he caught on a recent trip, and their sizes, and how he felt bad because one died before he could throw it back. His entries taper off after that, only return when he was hospitalized for the first time:

> *Wednesday, October 15, 2003* [nine days after his first tumor removal]
> *Donna and I went to Dr. M's office and visited the tumor.*
> *It was surreal to actually feel the cantaloupe sized mass in my hands, realizing it was growing in me for a long time.*
> *Cubs lost. I could not finish the game. I feel sorry for the city but for myself it had been a exciting ride and a good diversion during recovery.*

The diary entries span the dates of his hospitalizations over the years; as soon as he went home, he stopped writing. Thus, the diary is primarily an autopathography, a personal narrative of illness. He writes of the physical pain and, worse, the emotional pain that stems from what he saw as his uselessness as he lay in the hospital bed. ("I had always feared stasis more than I had feared pain," writes essayist Angie Chuang.)[4]

When I read through this journal, from beginning to end, I realized again how little I knew of my father's illness, so wrapped up as I was in my own. I was only privy to his mental state that one time when he told me he was going to the Elves, and even then, what could I offer him? Very little. In the diary I found his own perspective of that interaction: *Jami tried to help me by psychiatrizing me. I was appreciative, but not in the mood.*

In the car I flipped through the pages, searching for my name. He only mentioned our relationship in a couple other entries toward the beginning, before he got sick, when he still thought he'd write in the diary all the time. He wrote that I was at an age, fourteen, where we didn't spend a lot of time together or have much in common. A week later, he wrote:

> *I prayed for [a] better relationship with Jami and today I asked her what she would like to do with me . . . She would like to go to a show or a play. I think God is showing me to be intentional about my relationship with my girls and Donna.*

At the time he wrote this, I was crying to my friends at my church camp about my distant relationship with my parents. Now it struck me that we had worried over the same things and yet we never spoke of them. We were ciphers, each haunting the other while our bodies remained side by side. It would be another three years before he would sit on my bed, lift the covers, and ask me if I needed help.

I was able to read through the journal in one sitting, before my daughter woke from her nap. I'd thought I would cry. I didn't. He was still inscrutable to me, the way parents are to their children. I looked at my daughter, snoring softly in the back seat, head resting on her shoulder: she, too, would one day be inscrutable to me. How strange, I thought, that a body can come from yours and later be a foreign object. How inevitable.

———

Besides this journal, I have little other record of my father's handwriting. He did not write cards or letters to us; my mother was the communicator in our family. When I turned eighteen, she gave me a scrapbook tied with ribbon and filled with childhood photos, one page dedicated to each year of my life. On the last page, tucked in an envelope, were the letters she'd written me every birthday. The early letters—when I was one, when I was two—are on thin yellow sheets, typed on a typewriter. I read them and cried, to see myself in someone else's eyes, someone who loved me, rather than through the lens of my own biting hand.

(I am trying to replicate this tradition for my own daughter, but I am already delinquent. I worry about forgetting all of my daughter's quotes, worry I will forget who she was, though what is the point of creating the archive if I cannot bear to look at it?)

———

I wrote this book for my father, and I wrote it for me. Can both exist at once, can a text be both selfish and communal? Is there something between an exorcism and a resurrection? Else I will be stuck here forever, responsible only to the dead. (It is easy to probe—to interrogate—to love—the dead because in the end they do not require much of us except that we remember them; how much harder to love the living, who have wounds that we could actually attend.) Is this a narrative?

I am afraid of abandoning, which is to say—of abandonment.

While writing this book, I've become more and more isolated. The work siphons my emotional energy. I am looking forward to being a part of the world again. I keep telling people—*when I am done with the book*—but it seems an impossible thing, to ever be done, or for it to be done with me. (Though I tell myself a book and a body are not the same.)

I keep thinking if I look at the empty walls of my house long enough, someone will hang up these shelves.

———

Maybe I just need to wait for the text to turn one hundred so it can receive its own spirit, embark on its own journey. If my father were a tsukumogami he would be a fishing pole, its line sinking into the water. If I were a tsukumogami, I would be a blank book.

I do not have anything one hundred years old—only these stories, and few of those at that. The items that would be were left in other lands, or lost, or given up in incarceration. But in a couple decades the things I pass down to my daughter will reach this age. I will wait to see what happens when they come alive.

But one hundred years is so long to live. Every New Year's, my sisters and my husband and daughter and I gather at our mother's house for the traditional sukiyaki meal. We take turns pulling a shirataki noodle from the hot pot with our hashi, to see how long our noodle will be, and thus, how long our life. My mother always jokes: But I don't want *that* long a noodle.

Though maybe we change our minds as we age, as we see the generations that come after us. One autumn, my grandpa flew from California to visit me and my daughter, then his only great-grandchild. I took him to see my ama, for my mother's father and my father's mother had always gotten along. She had him test her new foot massager, she served us tea, they told me stories I had not heard before.

Before we left, my ama grabbed his shoulder and said: We will live to one hundred. My grandpa let out a deep belly laugh that spread his wrinkles far and wide.

I think we can do it, Ama insisted. As of this editing, he is ninety-four and she is ninety.

———

The very last line in my father's diary is from September 2018, a few months before he died.

There must have been a reason I was born, he writes, *why my mother gave birth to me.*

And then it ends. The final entries are filled with despair. That is the archive—my father's only handwritten text.

If a stranger read this notebook, they would think his death, and the previous fifteen years, contained only anguish. And yet I, his daughter, his imperfect chronicler, feel compelled to say: *but, but, but.* To point out the narrowness of the text against the years that he lived, and also, that he lived a life after he stopped keeping the diary.

Before he died he typed on his phone a list of advice for Cori and Kristi and me, which includes *Don't make travel plans while in the hospital under the influence of synthetic morphine* (a lesson he learned the hard way), *Don't take all the credit for your children, and don't take all the blame either*, and his number one piece of advice—he constantly had to remind himself of this, too—*People are more important than getting things done.*

The last thing he ever wrote to me: a message on the back of the wooden door hanger my mother handed to me on Christmas Day. His last words to us: *I love you, thank you for taking care of me.* His last action: squeezing my mother's hand.

His book is not the fullness of him. It is the truth, and it is a truth, and it is not the whole truth of his life, or his death.

I am trying to remember this for myself as well. It is hard. The kyōrinrin has long, extendable arms.

———

"Death is not only a common subject in Japanese folklore," write Michiko Iwasaka and Barre Toelken, "but seems indeed to be the principal topic in Japanese tradition; nearly every festival, every ritual, every custom is bound up in some way with relationships between the living and the dead, between the present family and its ancestors, between the present occupation and its forebearers."[5]

This is how we live.

My ama talks to her son every day. In my ama there is part of him that I still do not know, but that I might, through her, still come to know, through my questions and her stories.

Once, when we visited my Okinawan great-grandparents' graves in O'ahu, my grandma gave the family members incense to light. My grandfather wanted her to snuff it out; I think he thought it was too Buddhist a ritual for their Christian beliefs.

No, Tom, my grandmother said, as she handed out the incense. It was one of the only times I can remember her standing up to my grandfather in this way. We were standing by her parents' graves, and my grandfather relented.

———

The summer after his death, my family scattered some of my father's ashes in Colorado during a road trip to see Amache and to visit the hot springs. It had been a long year. We were tired, our bodies so tired, we thought the water might soothe us.

We had trouble finding an appropriate place to scatter the ashes. Eventually we stood in a parking lot in an observatory and threw them over a fence onto the mountainside. But there was no wind; the ashes did not scatter. They merely clumped onto the ground below.

We saved the rest to scatter at his favorite places—New Zealand, Japan—but with the pandemic and the state of the world, we do not know if or when those trips will happen.

Once upon a time, my father was just a man standing on the rocks of Hawaii scattering his own ashes—those fingernail clippings, those bits of sloughed-off skin—into the sea.

Is asking where the story ends as futile as asking where it begins?

———

Like my father, I believe in a God, a heaven; unlike my father, that belief doesn't make me less afraid. When he was dying my ghosts lurked in the middle of the night. Now they lurk in the daytime. Maybe I only know what haunts me in retrospect, the same way a person can meet a friend in the street—have a normal conversation—and then later find out that they'd encountered a yūrei. Maybe only in hindsight can you accurately name what scares you.

I question everything. What will happen? I do not know. What will happen? I do not know. God too is an uncanny thing, a thing I cannot explain—a yōkai. I believe because I believe, and this is not a satisfactory answer. The beginning and the end, the alpha and the omega, a circle. The snake eating its own tail.

Certainly, doubt has its positives: doubt as a way of seeking, as curiosity, or wondering. That is not why it plagues me. Doubt feels wrong in my body, this physical manifestation of my anxiety. I do not mind *not* knowing if, eventually, I can *know*. But so much of faith is being certain of what you will never see—at least not in this lifetime. It is a loss of control, one I feel in my chest. My opinions slide back and forth from day to day like an insect trapped in the windshield wipers.

This is why I imagine this text as a kind of book-length essay, a form suited for my constant uncertainty. In the essay there is freedom to wobble, to say a thing and bite it back. The essay is made for *in fact*s and *however*s, a form for the gun-shy, for commitment-duckers, for those who swerve. Essays are inherently political, Leslie Jamison posits, "because they are committed to instability."[6] (Though I wonder: How much instability can you commit to before you are called unstable, or committed?)

Once, while overcome with anxiety, I hid in a corner and hummed the hymn *Trust and obey, for there's no other way . . .* until my breathing returned to normal. A friend told me that, during a bad trip, he paced and chanted the Lord's Prayer until the effects of the mushrooms wore off.

These words can function as cries to God or only as rituals, remnants of what once comforted us.

As Clarice Lispector says, "Who has not asked himself at some time or other: Am I a monster or is this what it means to be a person?"[7] Perhaps the very certain, the very confident do not ask themselves this; perhaps they should. Perhaps the certainty that you are not the monster—that no matter what you do, you will never become the monster—is what gives rise to monstrous behavior. To never question yourself or the people you listen to. To never question your stories, or your ghosts, or your gods.

———

I am weary of the archive.

I need to do something else.

THE THREE CORPSES

—

Also known as the Three Worms, these creatures reside in
the human body, sucking out the life energy of their hosts.

THE THREE CORPSES

—

*All Taoist masters and physicians know how to
prescribe medicines to heal one's body, but they
do not know that the Three Corpses are crouching
within the human's abdomen, which are so potent
they make the prescriptions ineffective.*

—*TAISHANG LINGBAO WUFUXU*, CIRCA 400 C.E.[1]

When a human is born, the old lore says, three spirits enter the body. Depending on who you ask, these spirits are called the Three Corpses or the Three Worms.[2]

My cousin Lauren—who was born in Taiwan, whose parents are both Taiwanese, whose Taiwanese-ness is salient to her in a way that mine is not—was the one who taught me about these crouching corpses that take up residence in your heart, stomach, and head, and signify the past, present, and future.

Descriptions of these parasites vary from text to text. In a ninth-century illustration collected in a later version of the Daozang, or the Daoist Canon,[3] the first corpse looks like a muscular human leg with the horned head of a goatlike animal. The second corpse appears like the guardian lion-dog known as komainu in Japan and shisa in Okinawa. The third looks like an ancient official. All three hold scrolls.

But no matter their appearance, the Three Corpses have the same goals. To suck out their hosts' life energy the same way the black worm sucked out Haku the White Dragon's life in *Spirited Away*, a movie I return to again and again when I try to explain the world of yōkai and similar creatures to my friends.

The worms lie to us. They lull us with songs of the past and dreams of the future, and the more we listen, the more we veer from true consciousness. While we are distracted, the worms feed on us. They agitate illness within our bodies. They meticulously track our actions and report our sins to the gods. They welcome disease. They want us, the hosts, to die, for when we die, they will be free.

———

My tumor arrived one year after my father's tumors made their home in his stomach. Like a joke, mine first appeared during my baby shower.

You have something stuck in your teeth, my friend told me, gesturing toward my champagne flute filled with orange juice and fruit.

What she thought was a raspberry was the first stage of what I'd later learn is called a pyogenic granuloma of the gingiva. In layman's terms: a benign mouth tumor. When I pressed the dark red blister bubbling along my gumline, it bled and bled. It would not stop bleeding. For fifteen minutes of my baby shower, I stood in the bathroom as my friends decorated onesies with Sharpies. The rest of the day I alternated between holding ice and a clump of tissue to my gum; in all the photos from that day I am tight-lipped.

The tumor grew larger every day, expanding behind my teeth into a bulbous growth, soft and fleshy. Pyogenic granuloma can occur in up to 5 percent of pregnant women.[4] Our increased hormones make us susceptible, though I'd never heard of it before. Some stimuli—a local irritant, bacterial plaque—sparks the lesion, and from there the tumor grows rapidly as our hormones rise. If large enough, it must be surgically removed, though there's a chance it'll return later in the pregnancy. After pregnancy, it usually disappears on its own.

If my granuloma had appeared in a vacuum—that is, if my father hadn't also been dying of cancer—I would remember my tumor and its two removals only as a painful and deeply unpleasant part of a painful and deeply unpleasant pregnancy. Instead, my benign tumor gained outsize significance. Or rather: I imbued it with significance. When my father and I joked about our respective tumors, I latched onto the word *our*—holding on to the plural for as long as I could.

———

My father's first tumor was removed in 2003. Afterward, the story went, he kept it floating in a Mason jar on his desk. For years, I imagined a pink-tendrilled jellyfish suspended in clear liquid. I thought about it so often—what this piece of my father must look like—that it even played the starring role in one of my essays.

The story broke down slowly. Almost a decade and a half later, my father's surgeon showed us a photo of my father's twenty removed tumors arranged neatly on a tray, the largest the size of an orange, the smallest ones tinier than a blueberry, all red and pulpy—nothing like what I'd imagined.

And years after that, after my father is already dead, my mother tells me that he never kept his first tumor at all. He wasn't allowed; it had to be sent to Johns Hopkins or the Mayo Clinic or somewhere else for testing. Where did that story come from, then? I asked. She didn't know what I was talking about. What I had thought was family apocrypha was instead a story of my own imagining, re-encoded in my memory every time I rewrote it. In the end, I was talking only to myself.

(Later, when I am fact-checking this book, my mother remembers that my grandfather kept *his* growth in a jar after it was removed from his face. That must be what you got confused about, she says. The stories collapse.)

What is verifiably true, however, is that before the tumor departed his person, my father named it Tommy. Thereafter he referred to it as *him*. This is Tommy the Tumor, my father would say, the son I never had. To this my mother would roll her eyes in an exaggerated fashion and say *Charlie*; they'd been doing a variation on this bit their entire married life.

In the years between 2003 and 2017, when we thought the tumor was In the Past, I would laugh while repeating the Tommy tale to new acquaintances—*My bizarre family!*—squishing us into a box the size of that Mason jar.

———

In the two months between my baby shower and my daughter's birth, my tumor took up increasing space in my mouth and mind. Parasites have always terrified me, the idea of something living inside you, you but not you. How would you even know it was there? How could you prepare for an attack from within?

My father was the only one in the family who was willing to look at it as it developed. He possessed the certain fortitude necessary in a doctor—I did not—though we both loved the visceral and the grotesque. My first reaction: *Don't show me!* Then: *Show me.*

Granuloma: even the sound of the word is gross in the mouth, like bits of sand lodged between one's teeth.

My tumor grew so intrusive I had it removed by an oral surgeon, whose assistant asked me if she could take a photo, she had never seen a pyogenic granuloma this big. Later, doubled in pain—I was pregnant and could have none of the good pills—I could no longer feel the baby kicking. I thought perhaps the anesthesia had killed her. Only my father's voice on the phone could lull me back to logic.

After the removal, the tumor grew back even larger, soon reaching the size of a small grape. It looked like a knob of uncooked chicken breast limned with fat. Once, after dinner, I swallowed what I thought was a piece of meat trapped behind my teeth. Only afterward did I realize I'd chewed the edges of the tumor itself.

———

One of the Three Corpses is named the Bloody Corpse, and it is through this parasite that the host's "intestines are painfully twisted, that the bones are dried out, that the skin withers . . . that the will is not strong, that thought is infirm, that one can no longer eat and is hungry, that one is sad and sighs, that ardor flags, that the spirit wastes away and falls into confusion."[5]

My father and I moved at a similar pace then—him with a belly full of cancer, me with a belly full of baby. We sat on the couch together. We watched television. We talked about our tumors. I named my first mouth tumor Tammy, and when it returned, Tammy Two, after the vexing ex-wives of a main character from a sitcom I loved. The female version of my father's first Tommy, the one we believed was innocent.

He still named his new tumors, though we no longer thought them innocent. The biggest he called Malachi, after a character on *Longmire*, a show he and my mother checked out from the library one season after the other. After Malachi was removed, the next two were Ananias and Sapphira, biblical characters from the book of Acts who lie to the apostle Peter about how much they have donated to the church. They have secretly kept some of their wealth for themselves. *You have not lied to men but to the Holy Ghost*, Peter admonishes them. They are struck dead, first one, then the other.

Both of us were slow, cautious. But we were moving in different directions. My tumor died when my daughter was born. My father's died when he did.

————

According to folklore, to remove the Three Corpses you must make a concoction of root and liquor and flour and yeast and thicken it until it looks like caramel. Or you must make a tincture of cinnabar.⁶ Or you must use a special breathing technique while swallowing in a certain way. After thirty days you will excrete the lower corpse (the worm of your belly). After sixty days you will excrete the middle corpse (the worm of your heart). After one hundred days, you will excrete the superior corpse (the worm of your head). If you do this, you will be free. If you do this, you will not die.

The things he tried: debulking surgery, chemotherapy via pills, chemotherapy via infusion, radiation through his veins. And the tumors came back, and they came back, and they came back. Every week my ama drove over in her white Cadillac with food she thought could heal her son: purple potatoes, pudding cups, cooked broccoli in a plastic bowl.

But what is a purple potato against the tenacity of the Three Corpses seeking their own freedom? What is a broccoli floret against the Three Worms, who know that when their host dies, they may roam the world as ghosts, eating and drinking from family altars as they please?

————

"The peach and plum trees say nothing," says the fourteenth-century Japanese monk Yoshida Kenkō. "With whom is one to reminisce about the past?"⁷

After my father died, I feared the worm of the heart. I feared that I would be lulled by the siren song of the past and remain there, calcifying, the titular character in Faulkner's story who refuses to give up her father's corpse.

How seductive to remain in a life already lived, where the outcome is known, and not the one unfolding in its uncertain paths.

("[The] pull between past and present is one of the forces in the creation and re-creation of yōkai through time," writes Michael Dylan Foster.[8])

This fall I go to the dentist after avoiding it for years, fearful of the needle and traumatized from my own tumor surgeries. This time I am prepared: I take my Xanax and request the nitrous oxide, and when the needle jabs into my gums, I feel it only a little.

If I ended this story here, it would provide a moment of hope: grief transforming, life moving on. Instead, I'll tell you this: The dentist cannot fill all my cavities. He shows me my X-rays. There is my tooth, cavernous and open. You need a root canal, he says.

Only later, when it begins to hurt, do I understand that the tooth that needs the root canal is the one where my mouth tumor grew and grew and grew. A cavity is an absence. Nature abhors a vacuum.

What do we cut out? What do we let grow within us?

For so long before he died, I held an anger at my father. One singular anger, the size of a walnut. Once I was small and so sad and he couldn't see it. I was young and he was a doctor, and he was my father, and who was I to argue. Years later, after treatment, I thought, All that time would have been better *if*. And I held the walnut in my fist. I was angry with my father, though not for ignoring me. For being wrong. I could not let it go until I wrote a story about it and gave it to him for Father's Day. He read it and said, Thank you for finally telling me.

For so long after he died, anytime I wrote of my father he was not there. Only grief, only the yūrei in the shape of my father.

Here is my father's body: legs scaled with white lines, even when he was healthy, the driest of skin. Like a crocodile, he said. And on top, stray, wiry black hairs sticking straight out of the skin like spider's legs. From him I have inherited this skin, this hair. When I was a teenager, I used to lotion my legs vigorously to make the scales disappear, but my father did not care. After a while my legs dried up, too. I could no longer be bothered.

He was shorter than my mother: when I was young by half an inch or so, and by the time he was dying by a much more significant amount. Maybe it was because he was bald, and the hair added height. Maybe his spine was curling.

His skin a few shades darker than mine, and darker still in summer when he absorbed the sun.

The flat calluses on the knuckles of our toes.

Our anxiety.

His smile line that would travel up to his eyes like fissures in the earth.

His garrulous laugh.

———

There is a danger to cutting too much. Our bodies hold all that we are.

A couple of weeks before he died, my father invited a man from a crematorium to come to our house.

I want to pick out my urn, he explained as we waited for the man to arrive. If you wait until after I'm dead, they'll say something like, *Oh, Charlie would've wanted this expensive urn.* And then you'll all feel guilty and buy it. But they can't do that when I am sitting right here.

I'm going to say, *No, I want to be stored in this old yogurt container!* He gestured toward the half-empty Chobani sitting on the kitchen table.

He had done his research to ensure everything was orderly for his death and afterward. The cremation man arrived with his slick brochures and his attaché case and his spiel, but my father knew exactly what we wanted; all we had to do was sign the papers. I paid using my new credit card. My father repaid me right after. He just wanted to make sure I spent enough to get the bonus airline miles. If he had to die, he figured, I at least should get some free travel out of it.

(*I told the cremation people, if you don't take credit cards, I'm not dying with you*, he joked later, when retelling this story.)

After the man from the crematorium left, my father lay on the couch to rest. Those days, his energy had to be meted out by the teaspoon. He was wearing the heavy knit beige cardigan he wore at least every other day; his body had no fat to it, he was always cold, it was his cocoon.

Why cremation? I asked him.

That way my body can be one with the atoms, he said. I don't want it to be in the ground getting eaten by mice. After I trapped so many of their brothers, it would be their revenge.

He shifted into the cushions and continued, This way, he said, my body is everywhere.

———

Listen—

In the presence of a story, if the story is a good one, time collapses.

Though throughout the telling I have worried: *what is there to say, there is nothing new to say, death and grief are the oldest stories under the sun.*

But my horror is not of death, where the living and the ones they mourn are irrevocably dispatched to different timelines. It is of the death of memory.

For all my weariness of the archive I cannot rid myself of it, despite its psychic and literal space. As I write this, the boxes of papers are piled high in the corners of my bedroom. *What if we put this away?* my husband asks, and I say, *No.*

Perhaps in another timeline I will feel differently. I keep the archive so if anyone asks—if I ask myself—*Did this happen to you?* or *Was he loved?*, or if I forget, then there is this body that cannot be eaten by worms. *Yes,* and *yes.* I do not need it now, do not feel the need to look at it, but I know I will come back to it later. I hope I will have different questions then.

I may never be finished with the archive, though I must be finished with this book itself. When beginning this story, I wondered how I would know it was done. I thought I would feel a rapturous moment of completion: *It is over.* Instead, I know it is done because I can no longer sit in it without breaking. I am spent. It is October 2022, and my body and my mind are beginning to crack under the weight of this book, and I refuse to let them crack. I have learned this much. It still takes me too long to realize what my body is telling me, but I am listening. If this book is not perfect, then let it be enough, so I, too, can live outside of it.

My father, my body, the archive, this book—these are all different things. Let the last be finished so the others can breathe.

———

When the yōkai danced through the towns in the Night Parade, the residents would close their shutters and close their eyes. *Don't look, or you'll be spirited away.*

Thus, part of the pleasure of the scrolls that depicted the Night Parade was that you could look. The yōkai were "rendered visible, and gazed upon with pleasure," writes Michael Dylan Foster.[9] In these illustrations, "the mysterious spirits of untamed nature are transmuted into everyday objects; terror turns into humor."

The scrolls, and later yōkai encyclopedias like Toriyama Sekien's, allow us to see these creatures without putting our own bodies in danger. But Foster argues that the transformation—from something amorphous and terrifying to ordered and mundane—is not fixed, or one-way.

The yōkai move back and forth from one side to the other. Visible to invisible, legible to illegible. Terror turns into humor turns into terror.

Or as the narrator from a story by Lucia Berlin, one of my favorite writers, says: "I don't mind telling people awful things if I can make them funny."[10] Later, however, she admits: not everything is funny.

My father, of course, always told me I needed to add more jokes.

We look, we look away, we look again.

———

A book of the dead is for the living. A funeral, a memorial service, a grave—these are for the ancestors and those of us who will later become ancestors.

"According to Yanagita," scholars Michiko Iwasaka and Barre Toelken write, "*senzo* (ancestors) are not relatives who have vanished from the scene but personages who continue to exist *because* they are celebrated."[11]

A body is recomposed by the things it leaves behind.

———

From the beginning I knew that terror is a god. But now I also believe that what might sound like a death rattle is merely the echo of ancestral song. For there he is, singing a hymn off-key in the balcony of our church. There he is, whispering *Chou chou, my little chou chou*, in my newborn daughter's ear. There he is, holding my mother's hand as they walk in the woods behind their house. There he is, kissing our foreheads before we go to bed, calling us his precious treasures.

There he is, in the body of my ama, and in his sister and brother and their children, and in his in-laws and mentees and friends, and in my daughter, and in Kristi and Cori and me, and in my mother: in the body of all the people he ever loved. Whatever transformation happens now will occur in us.

There he is: all around us in the day, and in our dreams at night, and waiting in the place beyond.

There he is, sleeping on our green couch in his flannel pajamas. I watch the man who shouldered me all of my life, watch his hollow chest rise up and down as he breathes. Rest, Papu. Rest. You are so small now. Because you are so light, I can move your legs to sit next to you. Because you are so light, we can carry you always, wherever we need to go.

———

とぞいいつたえたる

To zo ītsutaetaru.

And so it has been passed on.

ACKNOWLEDGMENTS

This book and I would not exist without the support, love, and care of my large community.

Thank you to my family—to all the Iges and Changs and Nakamuras and Lins in this world and the next.

Thank you to Lisa Hofmann-Kuroda and Jenna Tang, the best translation and research team. The book is infinitely richer because I was not limited to the existing English translations of this folklore and research; thank you for being the vessel through which these stories could travel.

Thank you to Stephanie Delman, agent of agents, conjurer of dreams, for shepherding Cori and me through this entire process; I am so lucky you found me. Thank you also to Khalid McCalla, Michelle Brower, and the rest of the amazing team at Trellis Literary Management. (I also couldn't have gotten through this process without the tight-knit community and support of fellow Trellis clients!)

Thank you to my editor Jessica Williams for telling me to lean into the weird and getting the book to a place I never could have imagined. To Julia Elliott for all your support throughout this journey. I am also so grateful to the art team that met with us so many times and worked with Cori to create the art objects of our dreams—to Leah Carlson-Stanisic for the beautiful interior, and to Jeanne Reina for her art direction on the cover. Thank you also to Kasey Feather, Emily Fisher, Karen Richardson, Laura Brady, Pam Barricklow, Jennifer Hart, Liate Stehlik, Peter Hubbard, and everyone else at Mariner.

Thank you to the HarperCollins Union for fighting for a better, more equitable future for all of us in the industry.

Thank you to Molly Slight, Adam Howard, Sarah Braybrooke, and all the folks at Scribe UK.

Thank you to all the people with whom I was able to talk about these topics or who pointed me to further research, including Bruce Owens Grimm (whose "Haunted Memoir" essay in *Assay: A Journal of Nonfiction Studies* informed so much of my thinking), Shō Tanaka for the Okinawa section, Nicholas Fiorentino for the whales, H. Yumi Kim, Kiyoshi Mino, and Brandon Shimoda.

Several of the chapters in this book began their lives in other forms. Thank you to the folks at Catapult magazine (RIP) for giving my essay column "The Monster in the Mirror" a home. Without the column there would be no book. Thank you to my editor Leah Johnson, who saw the bones of each messy draft and went back and forth with me until the skeleton became a living, breathing thing. Thank you, too, to Matt Ortile and Nicole Chung. Thank you to Kao Kalia Yang and Shannon Gibney for editing the "The Night Parade" chapter and including it in the *What God Is Honored Here? Writings on Miscarriage and Infant Loss by and for Native Women and Women of Color* anthology. Thank you to Melonyce McAfee at the *New York Times* for accepting my piece about my daughter's name and to Nick Almeida Miller at *Bat City Review* for taking my very first musings about whales and motherhood.

Thank you to my writing community, whose friendship, group chats, and Zoom calls sustain me: aureleo sans, Bruce Owens Grimm, Jenny Tinghui Zhang, J. Marcelo (Mio) Borromeo, Nicole Zhao, Stephanie Cuepo Wobby, Tate Gieselmann, Temim Fruchter, Vanessa Chan, and Victor Yang. Thank you to Duende, Periplus Collective, and the Caucus. Thank you to my first cohort: Lauren Ostberg, Rebecca Kuensting, Abby Minor, Julie Johnson, and Max Vanderhyden.

Thanks to all these writers for their guidance, mentorship, teaching wisdom, or other forms of support: Aimee Bender, Alex Marzano-Lesnevich, Alexander Chee, Charlotte Holmes, Esmé Weijun Wang, Hugh "H.D." Hunter (in whose class I began to imagine "The Offing"), Jason Mott, Jaquira Diaz, Katie Kitamura, Kelly Link, K-Ming Chang, Maurice Carlos Ruffin, Sari Botton (in whose class I wrote "Skin, a Love Story"), and Sequoia Nagamatsu.

Thank you to David McGlynn, who asked a twenty-one-year-old if she had ever thought about getting an MFA and set me on this path, and to Mrs. Gilkey—wherever you are—who taught English 108 at Fremd High School and showed me what literature could do.

Thank you to the Japan-U.S. Friendship Commission and the National Endowment for the Arts for funding my research fellowship, and to the International House of Japan in Tokyo for hosting me. Thank you to Mihoko Sasei for your help with interpretation and for your friendship during that trip, and to all the other families and friends who supported me there. Thank you to Yaddo, Sewanee, the Sustainable Arts Foundation, Illinois Arts Council Agency, We Need Diverse Books, Wormfarm Institute, and Writers' Colony at Dairy Hollow for giving me the time and resources to write.

Thank you to Kundiman, VONA/Voices, FIYAH, Diverse Voices, and Asian American Writers' Workshop for all that you provide for our communities.

Finally—a body of work begins with a body, and mine wouldn't have made it this far without these people:

To my therapist Maggie, who probably now knows more about the writing and editing process than she ever cared to know. To our family therapist Kanchan, who got my sisters, my mother, and me through a hard year.

Thank you to Meg and all the Saikis for four generations of kinship. To the CCP community, especially the Monday-night and Friday-night women. To Lucy and Elise and Amy for all our years of friendship and motherhood. I am blessed to have too many friends and loved ones to thank by name, but I am grateful for all of you.

Thank you to all our cousins for modeling as the yōkai, to our ama for providing the brush artwork for the kanji, and to our grandpa for preserving so many of the memories.

Thank you to the God I do not understand.

Thank you to Cori for our lifelong collaborations, for the ways we are tangled and the ways we are untangling. For all the things I learn and grow from you, for all the things you create in the third and fourth dimensions. Thank you for helping me think through so many of these essays and for sparking the ideas in the first place.

To Kristi for all your years of love and silliness and wisdom. While your work might not be as visible in this book, Cori and I couldn't have made it without you.

To my mother for everything from the beginning. Thank you for helping me believe in transformation. I love you.

To Aaron, without whom nothing could be possible, and to C, who provides all the joy. It is the greatest blessing to spend each day with both of you.

To my father. I wish you could read this, but you know all it says.

Thank you to all those who came before me, all those who still are with me, and all those I cannot see anymore but still carry in my heart.

And last: Thank you, once again, to my family—the Iges, the Changs, the Nakamuras, and the Lins. Your names belong here, in the beginning and the end. It is not an easy road, but I am glad to walk it with you.

ILLUSTRATOR'S NOTE

———

Cori Nakamura Lin

My goal in illustrating *The Night Parade* was to depict yōkai that have escaped the idealized, frozen concept of ancient Japan and have re-rooted in the complexities of my multicultural Japanese American experience. I drew concepts and iconography stemming not only from historical renderings of yōkai—sources like the ancient Hyakki Yagyō Emaki picture scroll attributed to Tosa Mitsunobu; the eighteenth-century yōkai encyclopedias of Toriyama Sekien; the art exhibited in *Yōkai: Ghosts, Demons and Monsters of Japan*, edited by Felicia Katz-Harris—but also contemporary Japanese anime *GeGeGe no Kitarō, Yo-kai Watch*, Studio Ghibli, *Pokémon*, and Asian American media like *Usagi Yojimbo* and *Mulan*.

The backgrounds for the yōkai and other Taiwanese and Okinawan figures are based on settings from my family's migration path: islands in the East China Sea, the farmlands of California, the desert of Amache incarceration camp, the clear-cut old-growth forests of the Midwestern plains, Jami's backyard garden in the Chicago suburbs.

Each of the humanoid yōkai were modeled after Jami and my sister and our cousins, whose likenesses I gathered from screenshots and selfies taken in their dorms and homes across the continent. I chose to add symbols and artifacts from our mixed yonsei upbringing: a Locals zori, a purity ring, a Magic Eraser, a Nintendo 64 game system. Finally, I painted each chapter illustration in gouache and watercolor, then cut the paintings out with a scalpel. The kanji characters for kishōtenketsu that begin parts 1, 2, 3, and 4 were brush-painted by our ama—our father's mother—Toshiko Lin.

As an artist who lives and creates in the heartland of an empire, I wanted to visualize spirits that mourn war, displacement, and loss of cultural memory. The spirits I painted for *The Night Parade* evoke cycles, release rage, channel joy, embody in-betweenness, and remember what's been lost. I hope my visions of yōkai and spirits who have traveled across occupied land as well as time will inspire other diasporic people to face and heal the ghosts that haunt us.

THANKS:

With all my gratitude and love to Jami for being the words to my visions, my first collaborator, my first believer.

To my mom for inspiring me to be a maker. To Kristi for being the unnamed fulcrum of this creative team. And to my father, in the spirit realm, who showed me how to love the ocean.

To the other members of the Bi Council: Tori Nxtoo Hong and Christian Aldana, whose mixed queer brilliance created a haven for me in the between-space.

To my doushi in Nikkei Uprising, whose care and commitment remind me to yearn toward a japaneseness that doesn't yet exist.

To Mizuki Shigeru's nanny, NonNonBa, who preserved the precious stories before we knew they were precious. To my ima, Shirley Chang (d. 2021), whose inheritances give me the stability to dream. To Taira Toshiko (d. 2022) and Tomiyama Taeko (d. 2021), whose legacies remind me that one never creates alone.

And to the original land stewards, who taught me that the land is a spirit. With reverence to the islands, the ocean, my ancestors, and the future generations.

ILLUSTRATION REFERENCES

Many of the sources stem from Toriyama Sekien's work collected in *Japandemonium Illustrated: The Yokai Encyclopedias of Toriyama Sekien*, translated by Hiroko Yoda and Matt Alt (hereafter referred to as *Japandemonium*), and *Yōkai: Ghosts, Demons and Monsters of Japan*, edited by Felicia Katz-Harris (hereafter referred to as *Yōkai*).

THE DRAGON KING: RYŪJIN

* This illustration is inspired by many historical depictions of the Dragon King, Mazu the Sea Goddess, and Haku from Hayao Miyazaki's *Spirited Away* movie.

AUTOMYTHOLOGIES I: KAPPA

* The "old" kappa (the kappa of folklore) is modeled on our cousin Ethan Lin. The illustration is inspired by Toriyama's version of the kappa

(*Japandemonium*, page 16) and his version of the suiko, the Chinese interpretation of the kappa (*Japandemonium*, page 91).

+ The "new" kappa (the kappa of folklorism) is based on our cousin Taylor Lin Okamoto. The illustration is inspired by Mizuki Shigeru's kappa from the manga *GeGeGe no Kitarō*, the photos Jami took of Tōno village mascots and statuary, and the Sumida City village mascot.

THE RAGE: ONI-BABA

+ This is a self-portrait. The illustration is inspired by Toriyama's oshiroi-baba (*Japandemonium*, page 176) and yamauba (*Japandemonium*, page 50).

THE TEMPLE OF THE HOLY GHOST: ROKUROKUBI

+ This illustration is modeled on our cousins Lauren and Megan Lin. The setting is a teen girl's bedroom. The illustration is inspired by Toriyama's rokurokubi (*Japandemonium*, page 176), the "Obake karuta" playing cards (depicted in *Yōkai*, circa 1860, artist unknown, page 177), and the nukekubi from the *Bakemono-zukushi* (Edo period, artist unknown).

THE OFFING: BAKEKUJIRA

+ This setting is our cultural center—Japan, Taiwan, Okinawa—the series of islands close together. The illustration is inspired by Mizuki's bakekujira and Hannako Lambert's bakekujira (Instagram: @thisishannako).

POSSESSION: KITSUNE

+ This illustration is modeled on my sister Jami and her daughter in their garden. Jami is wearing a sweater that she knit depicting Totoro, and her daughter is wearing a shirt inspired by Hello Kitty. The illustration is inspired by the 1852 woodblock print "Tsumagome (Abe no Yasune Watching His Wife Change into a Fox-Spirit)" by Utagawa Kuniyoshi, Toriyama's kitsune-bi (*Japandemonium*, page 22), and Benjamin Lacombe's 2020 illustration depicted on the cover of Lafcadio Hearn's book *Esprits & Créatures du Japon*.

SKIN, A LOVE STORY: SNAKE

+ The outside of the circle is based on decorative Chinese patterns. The illustration of the snake itself is based on the serpent yōkai uwabami and the Taiwanese coral snake.

THE TWILIGHT HOURS: DUGONG

+ This illustration is based on the zan, the Okinawan dugong, who are matrilineal, and shisa statues from Okinawa.

THE NIGHT PARADE: HYAKKI YAGYŌ

+ This illustration as a whole was inspired by the Hyakki Yagyō Emaki picture scroll attributed to Tosa Mitsunobu, drawn in the Muromachi period (1336–1573), and other picture scrolls. *Clockwise from top:* The blanket yōkai was inspired by Mitsunobu. The futakuchi-onna (with the mouth in the back of the head) is modeled on my sister Jami and inspired by the illustration in Takehara Shunsensai's *Ehon Hyaku Monogatari* (circa 1841). The yōkai with the black hat is inspired by Mitsunobu. The nurarihyon was inspired by Toriyama. The yōkai with the big nose and the umbrella mask are inspired by Mitsunobu. The jami (spirit yōkai) is inspired by Mizuki's *GeGeGe no Kitarō* version and Toriyama. I also added a lost earbud—a yōkai of the twenty-first century—and what I call "Daikon Pubes Yōkai," my interpretation of the daikon spirit in Miyazaki's *Spirited Away* film.

THE HOUR OF THE OX: YŪREI

+ This illustration is modeled on our sister Kristi Lin. The illustration is based on the hanging scroll "Female ghost kakejiku" (*Yōkai*, artist unknown, from the early twentieth century, page 27), Toyohara Kunichika's 1884 triptych (*Yōkai*, page 114), and Toriyama's version (*Japandemonium*, page 50).

MOURNINGTIME: BAKU

+ This illustration was inspired by Mizuki's baku, baku carvings from the early Edo period (seventeenth century), and general images of tapirs. The trees are Illinois white cedars (old growth forest), and the scene is a reference to Robin Wall Kimmerer's experience with clear-cutting from her book *Gathering Moss*.

THE NAMING: AMABIE

+ This illustration is modeled on our cousin Lia Camper. The illustration is inspired in part by artist Kiyoshi Mino's felt amabie that he created for our Yōkai Banzai event and the original 1846 woodblock print (artist unknown) that got passed down the post road.

THE YEAR OF THE RAT: RAT

+ The outside of the circle is based on decorative Chinese zodiac patterns. The nest includes bits of money—old and new U.S. dollars, Taiwanese yen under Japanese occupation, and the New Taiwan dollar. The glasses the rats are wearing represent our parents' glasses when we were growing up, including our father's tinted aviators.

IN THE WHIRLPOOLS: ONI

+ This illustration is modeled on our cousins Kenna Hadassah Camper and James Ko Nakamura Camper and set in Amache (Granada) incarceration camp. Kenna is holding a kanabō, a spiked two-handed war club (wielded by both oni and samurai!). I wanted to emphasize the land of Amache, focusing on the Amache gardeners. The illustration was inspired by the 2017 oni otsu-e art of Shozan Takahashi IV, Toriyama's oni (*Japandemonium*, page 88), and Edo-period oni, including the one from *Jigoku-zoshi Emaki* (depicted in *Yōkai*, page 80).

AUTOMYTHOLOGIES II (REPRISE): TSUKUMOGAMI

+ This illustration was inspired by the tsukumogami (tool yōkai) historical picture scrolls, including Mitsunobu's and Sakyu's "Modern Hyakki Yagyō" 2018 digital illustrations (depicted in *Yōkai*, page 237).

+ Jami and I also wanted to incorporate both old and new tools—not just tools from old Japan but also those from our yonsei Japanese American life in Chicago. *Clockwise from top right:* The big yōkai is a modern version of the kyōrinrin, the yōkai of the archive. Originally made out of sutras, here it is transformed and made out Jami's journals with a three-ring binder for a mouth. Behind it is a Locals zori like the ones our grandparents wear. The teapot and the fishing pole represent our mother's ceramics and our father's favorite hobby. The trowel represents our ama, who loves gardening, and behind it is a Nintendo 64 system. Finally, a Magic Eraser.

THE THREE CORPSES: SANSHI

+ The left corpse—a human foot with a cow's head—is modeled on our cousin Jay Nakamura Clark. The middle corpse—an animal— is modeled on our aunt Debbie Shimabukuro Cast's dog Ernie. The right corpse—a scholar—is modeled on our cousin Kyle Shimabukuro Smith. The setting is our parents' suburban home, with our father on the green couch. The illustration is inspired by the ancient Chinese image of the Three Corpses, believed to be from between the 600s and the 900s.

NOTES

PART I: KI

1. "If we rephrase kishōtenketsu so that it is easily understandable to children, we might call the four parts 'beginning' (hajime), 'continuation' (tsuzuki), 'change' (kawari) and 'ending' (musubi—literally, tying up loose ends)." Imai Takajirō, *Sakubun no Jugyō Nyūmon* (Tokyo: Meiji Tosho Shuppan, 1961), 187, trans. Lisa Hofmann-Kuroda.
2. Suzuki Toshio, "Yōsai no 'Kishōtenketsu' Sesshakujun," *Tōyō Shihō* 3 (1997): 8–18, trans. Lisa Hofmann-Kuroda. Yōsai is the Japanese version of Yang Zhai's name.

CHAPTER 1: THE DRAGON KING

1. This is transformed from several versions of Urashima Tarō I've read over the years, including the book I first encountered in the library as a child—Florence Sakade, *Urashima Tarō and Other Japanese Children's Stories* (Rutland, VT: Tuttle Publishing, 1986)—and in Ozaki Yei Theodora, *The Japanese Fairy Book* (New York: Dover, [1903] 1967), 26–42. Though Ozaki's name is the only one printed on the cover ("compiled by"), in the preface she notes that this book is a loose translation of work by Sadanami Sanjin (the pseudonym of Iwaya Sazanami, 1870–1933), a Japanese novelist and proponent of children's literature. There are many different versions of the story both in Japan and Taiwan.
2. Allan G. Grapard, "The Shinto of Yoshida Kanetomo," *Monumenta Nipponica* 47, no. 1 (1992): 30, https://doi.org/10.2307/2385357.
3. Florence Sakade, *Japanese Children's Favorite Stories* (North Clarendon, VT: Tuttle Publishing, 2014), 80.
4. From the historical tale *Mizukagami*, written around 1195, as described in Marinus Willem de Visser, *The Dragon in China and Japan* (Amsterdam: J. Müller, 1913), 142–43.
5. Michiko Iwasaka and Barre Toelken, *Ghosts and the Japanese* (Logan: Utah State University Press, 1994), 103.
6. For more on the traditions and worship rituals of the fishing and coastal villages, see Mayumi Itoh, *The Japanese Culture of Mourning Whales: Whale Graves and Memorial Monuments in Japan* (London: Palgrave Macmillan, 2018).
7. As described in Basil Hall Chamberlain, trans., *The Kojiki: Records of Ancient Matters* (North Clarendon, VT: Tuttle Publishing, [1882] 1981), 145–47, and in F. Hadland Davis, *Myths and Legends of Japan* (Mineola, NY: Dover, [1913] 1992), 91.

CHAPTER 2: AUTOMYTHOLOGIES I

1. Komatsu Kazuhiko, "What Is a Yōkai," trans. Kaya Laterman and Satori Murata, in *Yōkai: Ghosts, Demons and Monsters of Japan*, ed. Felicia Katz-Harris (Santa Fe: Museum of New Mexico Press, 2019), 55.
2. My understanding of yōkai was greatly shaped by the work listed above, in addition to: Komatsu Kazuhiko, *An Introduction to Yōkai Culture*, trans. Hiroko Yoda and Matt Alt (Tokyo: Japan Publishing Industry Foundation for Culture, 2017); Toriyama Sekien, *Japandemonium Illustrated: The Yokai Encyclopedias of Toriyama Sekien*, trans. Hiroko Yoda and Matt Alt (New York: Dover, 2017); Michael Dylan Foster, *The Book of Yōkai: Mysterious Creatures of Japanese Folklore* (Oakland: University of California Press, 2015); and Foster, *Pandemonium and Parade* (Berkeley: University of California Press, 2009).
3. Foster, *Book of Yōkai*, 14.
4. Komatsu, *Yōkai Culture*, 81.
5. Yanagita Kunio and Sasaki Kizen, *Folk Legends from Tōno: Japan's Spirits, Deities, and Phantastic Creatures*, trans. Ronald Morse (Lanham, MD: Rowman & Littlefield, 2015), 37.
6. Yanagita and Sasaki, *Folk Legends from Tōno*, 37.

7. Noriko T. Reider, *Seven Demon Stories from Medieval Japan* (Boulder: University Press of Colorado, 2016), 111–14.
8. Noriko T. Reider, "The Appeal of 'Kaidan,' Tales of the Strange," *Asian Folklore Studies*, 59, no. 2 (2000): 267.
9. Komatsu, *Yōkai Culture*, 67–70.
10. Through Inoue Enryō, "we find the first systematic and rigorous attempt to exorcise the spirits from the Japanese landscape." Foster, *Pandemonium and Parade*, 77.
11. Yanagita Kunio, "About Folk Tales," in *The Yanagita Kunio Guide to the Japanese Folktale*, trans. Fanny Hagin Mayer (Bloomington: Indiana University Press, 1986), https://publish.iupress.indiana.edu/projects/the-yanagita-kunio-guide-to-the-japanese-folk-tale.
12. Larissa Pham, *Pop Song: Adventures in Art and Intimacy* (New York: Catapult, 2021), 51.
13. Diana I. Tamir et al., "Media Usage Diminishes Memory for Experiences," *Journal of Experimental Social Psychology* 76 (2018): 161.
14. Susan Sontag, *Illness as Metaphor and AIDS and Its Metaphors* (New York: Anchor, 1990), 3.
15. Carmen Maria Machado, *In the Dream House: A Memoir* (Minneapolis: Graywolf Press, 2019), 5.
16. Yanagita, "About Folk Tales," *Guide to the Japanese Folktale*.
17. Foster, *Pandemonium and Parade*, 139.
18. As quoted in Joan Acocella, "Once Upon a Time: The Lure of the Fairy Tale," *The New Yorker*, July 16, 2012, https://www.newyorker.com/magazine/2012/07/23/once-upon-a-time-3.
19. Marilyn Ivy, *Discourses of the Vanishing: Modernity, Phantasm, Japan* (Chicago: University of Chicago Press, 1995), 70–73.
20. Gerald Figal, *Civilization and Monsters: Spirits of Modernity in Meiji Japan* (Durham, NC: Duke University Press, 2000), 14.
21. Ibid., 15.
22. Yanagita did, however, push back against some Meiji-era establishment scholars who both (a) discredited the importance of the stories of ordinary people and (b) ignored how Japan's folklore traditions had been influenced by those of China and Korea. For more, see Michiko Iwasaka and Barre Toelken, *Ghosts and the Japanese* (Logan: Utah State University Press, 1994), 49–57.
23. Michael Dylan Foster, "The Metamorphosis of the Kappa: Transformation of Folklore to Folklorism in Japan," *Asian Folklore Studies* 57, no. 1 (1998): 14.
24. Betsy Pickle, "Al-Anon Helps Family, Friends to Orderly Lives," *Knoxville News-Sentinel*, October 11, 1981.
25. Yanagita and Sasaki, *Folk Legends from Tōno*, x.
26. Komatsu, *Yōkai Culture*, 79.
27. Yanagita, "About Folk Tales," *Guide to the Japanese Folktale*.
28. Walter Benjamin, "Storytelling and Healing," in *The Storyteller Essays*, trans. Tess Lewis (New York: NYRB Classics, 2019), Kindle.

PART 2: SHŌ

1. Imai Takajirō, *Sakubun no Jugyo Nyumon* (Tokyo: Meiji Tosho Shuppan, 1961), 187, trans. Lisa Hofmann-Kuroda.
2. Suzuki Toshio, "Yōsai no 'Kishōtenketsu' Sesshakujun," *Tōyō Shihō* 3 (1997): 8–18, trans. Lisa Hofmann-Kuroda.

CHAPTER 3: THE RAGE

1. Story transformed from several translations of *Kankyō no Tomo (A Companion in Solitude)*, written in 1222 by Keisei, including those found in the following: Rajyashree Pandey, "Women, Sexuality, and Enlightenment: Kankyo No Tomo," *Monumenta Nipponica* 50, no. 3 (1995): 325–56, https://doi.org/10.2307/2385548, and Keisei, "A Companion in Solitude," trans. Michael Emmerich, in *Traditional Japanese Literature: An Anthology, Beginnings to 1600*, ed. Haruo Shirane (New York: Columbia University Press, 2007), 292–94.
2. Eduard Vieta et al., "Protocol for the Management of Psychiatric Patients with Psychomotor Agitation," *BMC Psychiatry* 17, no. 1 (August 2017), https://doi.org/10.1186/s12888-017-1490-0.
3. Lana Burgess, "What Is Psychomotor Agitation?" Medical News Today, Healthline Media, March 9, 2022, https://www.medicalnewstoday.com/articles/319711.
4. Keisei, "A Companion in Solitude," 294.
5. Pandey, "Women, Sexuality, and Enlightenment," 340.
6. Keisei, "A Companion in Solitude," 292.
7. Mayako Murai, *From Dog Bridegroom to Wolf Girl: Contemporary Japanese Fairy-Tale Adaptations in Conversation with the West* (Detroit: Wayne State University Press, 2015), 21. In her translation of Baba Akiko, Murai translates the word *oni* as "demon"; I prefer to keep the word *oni*.

8.   Murai, *From Dog Bridegroom to Wolf Girl*, 21.
9.   Pandey, "Women, Sexuality, and Enlightenment," 340.
10. Charles Dickens, *A Tale of Two Cities* (New York: Bantam Classics, 2003), 283.

CHAPTER 4: THE TEMPLE OF THE HOLY GHOST

1.   1 Corinthians 6:19.
2.   This trait is found throughout several stories, and also listed in Hiroko Yoda and Matt Alt, *Yōkai Attack!* (North Clarendon, VT: Tuttle Publishing, 2012), 144.
3.   Story adapted from several translations of the tale originally told in the *Kasshi Yawa* by Matsura Seizan around 1821, found on various folktale and monster forums and wikis.
4.   My version is adapted from the *Kaibutsu Yoron* (Sasama Yoshihiko, 1994) as recounted in *Nihon Mikakunin Seibutsu Jiten*, ed. Sasama Yoshihiko (Tokyo: Kadokawa, 2018), Kindle, trans. Lisa Hofmann-Kuroda; and Lafcadio Hearn's version of the same tale, translated into English, in Lafcadio Hearn, *Kwaidan: Stories and Studies of Strange Things* (North Clarendon, VT: Tuttle Publishing, [1904] 1971).
5.   This dialogue is quoted directly from the version recounted by Hearn, *Kwaidan*, 92.
6.   Bessel A. van der Kolk, *The Body Keeps the Score: Brain, Mind, and Body in the Healing of Trauma* (New York: Penguin Books, 2015), chap. 4, Kindle.
7.   Shibata Shōkyoku, ed., *Kidan Ibun Jiten* (Tokyo: Chikuma Shobō, 2008), 704–5, trans. Lisa Hofmann-Kuroda.
8.   Shūji Tomita, *Ehagaki de Miru Nihon Kindai* (Tokyo: Seikyusha, 2005), 131, trans. Lisa Hofmann-Kuroda.

CHAPTER 5: THE OFFING

1.   Liao Hung-chi, "Stranded," trans. Jacqueline Li, *The Willowherb Review* (December 2021), https://www.thewillowherbreview.com/stranded-liao-hungchi-.
2.   This story is very loosely transformed from several versions of the bakekujira tale including the ones found in Mizuki Shigeru, *Mizuki Shigeru no Sekai Genjū Jiten* (Tokyo: Asahi Shinbunsha, 1994), trans. Lisa Hofmann-Kuroda, and in Matthew Meyer, *The Night Parade of One Hundred Demons: A Field Guide to Japanese Yōkai*, self-published, 2015.
3.   This story is transformed from Matsuzaki Kenzō, "Yorikujira no shochi wo megutte: Dōshokubutsu no kuyō," *Nihon Jōmin bunka kiyō* 19 (1996): 58, trans. Lisa Hofmann-Kuroda.
4.   "Whale Explodes in a Taiwanese City," BBC News, January 29, 2004, http://news.bbc.co.uk/2/hi/science/nature/3437455.stm.
5.   Dan Bloom, "'There She Blows!'," *Daily Mail*, November 27, 2013, https://www.dailymail.co.uk/news/article-2514317/Horrifying-footage-shows-washed-sperm-whale-EXPLODING-biologist-tries-cut-carcass.html.
6.   The video was originally captured by the Faroese Broadcasting Corporation and can be viewed here: http://kvf.fo/netvarp/sv/2013/11/26/video-her-brestur-hvalurin.
7.   Alan Taylor, "5 Years Since the 2011 Great East Japan Earthquake," *The Atlantic*, March 10, 2016, https://www.theatlantic.com/photo/2016/03/5-years-since-the-2011-great-east-japan-earthquake/473211.
8.   Thomas E. Ferrari, "Cetacean Beachings Correlate with Geomagnetic Disturbances in Earth's Magnetosphere: An Example of How Astronomical Changes Impact the Future of Life," *International Journal of Astrobiology* 16, no. 2 (2017): 163–75, https://doi.org/10.1017/S1473550401600252.
9.   Zack Davisson, "Bakekujira and Japan's Whale Cults," *Hyakumonogatari Kaidankai*, May 10, 2015, https://hyakumonogatari.com/2013/05/10/bakekujira-and-japans-whale-cults/.
10. Mayumi Itoh, *The Japanese Culture of Mourning Whales: Whale Graves and Memorial Monuments in Japan* (London: Palgrave Macmillan, 2018), 3.

CHAPTER 6: POSSESSION

1.   Lafcadio Hearn, *Glimpses of Unfamiliar Japan* (Tokyo: Tuttle Publishing, [1894] 2009), 259.
2.   This idea came up repeatedly in the work I read on fox possession, but the chapter "Fox Spirits in Villages" in H. Yumi Kim's *Madness in the Family: Women, Care, and Illness in Japan* (New York: Oxford University Press, 2022) provides a particularly detailed look.
3.   Matthew Meyer, *The Book of the Hakutaku*, self-published, 2018, 208.
4.   Matthew Salesses, *Craft in the Real World: Rethinking Fiction Writing and Workshopping* (New York: Catapult, 2021), 19.
5.   Kim, *Madness in the Family*, 28.
6.   Matthew Meyer, "Kyubi no Kitsune," *A Yōkai a Day*, October 7, 2007, https://matthewmeyer.net/blog/2009/10/07/a-yōkai-a-day-kyubi-no-kitsune.

7. For more on the Hyakki Yagyō as a collective noun, see Michael Dylan Foster, *Pandemonium and Parade* (Berkeley: University of California Press, 2009), 9.

8. Michael Bathgate, *The Fox's Craft in Japanese Religion and Culture: Shapeshifters, Transformations, and Duplicities* (New York: Routledge, 2004), xi.

9. Iwao Hino, *Dōbutsu Yōkai Tan: Volume 1* (Tokyo: Chuko Koron Shinsha, [1926] 2006), 76, trans. Lisa Hofmann-Kuroda.

10. Zack Davisson, "Kitsune no Yomeiri–The Fox Wedding," *Hyakumonogatari Kaidankai*, July 19, 2013, https://hyakumonogatari.com/2013/07/19/kitsune-no-yomeiri-the-fox-wedding/.

CHAPTER 7: SKIN, A LOVE STORY

1. Juan Eduardo Cirlot, *A Dictionary of Symbols*, trans. Jack Sage and Valerie Miles (New York: NYRB Books, 2020), 389.

2. Yanagita Kunio and Sasaki Kizen, *Folk Legends from Tōno: Japan's Spirits, Deities, and Phantastic Creatures*, trans. Ronald Morse (Lanham, MD: Rowman & Littlefield, 2015), 215–32.

3. Ibid., 101.

4. I have transformed this story from several different versions, including Jenna Tang's own oral retellings and her translations of the following sources: Chou Yu-Lan, Ding Yi-Tsai, and the Committee of Chinese Classical Literature, *Bei She Chuan Tong Su Ben* [The Popular White Snake Legend] (Taipei: Silkbook, 2007), and Yun Ying Tian Guang, "Brief Analysis of Different Versions of White Snake Legend (The Origin of the White Snake): A Review," *Weiwenku*, January 20, 2019, https://www.gushiciku.cn/dc_dr/KGV6.

5. Chou, Ding, and the Committee of Chinese Classical Literature, *Bei She Chuan Tong Su Ben*, 188–89.

PART 3: TEN

1. Imai Takajirō, *Sakubun no Jugyō Nyūmon* (Tokyo: Meiji Tosho Shuppan,1961), 187, trans. Lisa Hofmann-Kuroda .

2. Suzuki Toshio, "Yōsai no 'Kishōtenketsu' Sesshakujun," *Tōyō Shihō* 3 (1997): 8–18, trans. Lisa Hofmann-Kuroda.

CHAPTER 8: THE TWILIGHT HOURS

1. From an interview between folklorist Tanigawa Kenichi and scholar Ōe Osamu, published in *Tamashī no Minzokugaku: Tanigawa Kenichi no sekai* (Tokyo: Tomiyamabō Intanashonaru, 2006), from a section of the book called "What Is a God?" (Kami to wa nani ka), trans. Lisa Hofmann-Kuroda.

2. Uchida Senzō, "Basic Information Regarding a Rare and Wild Aquatic Creature of Japan," Ministry of Agriculture, Forestry and Fishing, Fisheries Agency Volume, Japanese Fisheries Agency, 1994, 569–83, trans. Lisa Hofmann-Kuroda.

3. Shimabukuro Genshichi, "Okinawa ni okeru yorimono," *Minzoku Denshō* 15, no. 11 (1951): 9, trans. Lisa Hofmann-Kuroda.

4. United Nations Environment Programme, *Dugong Status Report and Action Plans for Countries and Territories*, ed. Helene Marsh, 2002, 52, https://portals.iucn.org/library/sites/library/files/documents/2002-001.pdf.

5. I recommend the short film *Our Island's Treasure*, by Okinawan American high school student Kaiya Yonamine. She interviews many Okinawan elders who are protesting the building of the base and who explain how integral the ocean is to their life, community, and culture. https://vimeo.com/340517922.

6. Maia Hibbett, "In Okinawa, the US Military Seeks a Base Built on the Bones of the War Dead," *The Nation*, February 18, 2021, https://www.thenation.com/article/world/japan-okinawa-henoko/.

7. Matthew Meyer, *The Hour of Meeting Evil Spirits: An Encyclopedia of Mononoke and Magic*, self-published, 2015, 14.

8. Komatsu Kazuhiko, *An Introduction to Yōkai Culture*, trans. Hiroko Yoda and Matt Alt (Tokyo: Japan Publishing Industry Foundation for Culture, 2017), 167.

9. Komatsu, based on the ideas of Akasaka Norio, *Yōkai Culture*, 175.

10. Uesedo Tōru, *Taketomi Tōshi Minwa* (Tokyo: Hōsei Daigaku Shuppan kyoku, 1976), trans. Lisa Hofmann-Kuroda.

11. Like in many situations of mass casualty, it is difficult to pinpoint the exact number of deaths. Some American sources list the number of Okinawan deaths as low as one hundred thousand, but I am using the number provided by Okinawa Prefecture itself, which includes people who died at the hand of American troops, those who died by suicide, and those who were never found. "Number of Names Inscribed," Okinawa Prefecture, June 24, 2022, https://www.pref.okinawa.jp/site/kodomo/heiwadanjo/heiwa/7812.html, trans. Lisa Hofmann-Kuroda.

12. Elizabeth Miki Brina, *Speak, Okinawa: A Memoir* (New York: Knopf, 2021), Kindle.
13. United Nations Environment Programme, *Dugong Status Report*, 47.
14. Statistic from the Save the Dugong Campaign, as referenced in Miyume Tanji, *Myth, Protest and Struggle in Okinawa* (New York: Routledge, 2006), 205.
15. "With a Population Under a Dozen, Okinawa Dugongs Labeled Critically Endangered' by IUCN," *Japan Times*, December 12, 2019, https://www.japantimes.co.jp/news/2019/12/12/national/science-health/okinawa-dugongs-critically-endangered.

CHAPTER 9: THE NIGHT PARADE

1. Genesis 18:13–15.
2. I based my description on what is believed to be one of the oldest Hyakki Yagyō Emaki (picture scroll)—the sixteenth-century *Hyakki Yagyō Zu*, held at Shinjuan temple in Kyoto, and digitally archived through the University of Chicago: https://scrolls.uchicago.edu/view-scroll/7.
3. I first encountered the futakuchi-onna in Matthew Meyer, *The Night Parade of One Hundred Demons: A Field Guide to Japanese Yōkai*, self-published, 2015, 128.
4. Tale adapted from Takehara Shunsensai's 1841 bestiary, *Ehon hyaku monogatari*, as recounted in Murakami Kenji, *Yōkai Jiten* (Tokyo: Mainichi Shinbunsha, 2013), Kindle, trans. Lisa Hofmann-Kuroda.
5. Information on the hone-onna was gathered from Toriyama Sekien's *Konjaku Gazu Zoku Hyakki* (as translated in Toriyama, *Japandemonium Illustrated: The Yokai Encyclopedias of Toriyama Sekien*, trans. Hiroko Yoda and Matt Alt [New York: Dover, 2017], 126), and Meyer, *The Night Parade*, 134.
6. Hanya Yanagihara, *A Little Life* (New York: Doubleday, 2015), 164.
7. Philippians 4:6–7.
8. Genesis 30:1–3.
9. Sources consulted include Yanagita's *Legends of Tōno* and Zack Davisson's introductory essay from his translation (*Tōno monogatari*, Montreal: Drawn & Quarterly, 2021) of Mizuki Shigeru's 2010 graphic novel based on Yanagita's work. Yanagita, of course, heard the tales from Sasaki Kizen, who heard the tales from locals. Thus, I retell this story fifth-hand.
10. Toriyama, *Japandemonium*, 134.
11. Ibid.
12. The history and background of the nurarihyon is riddled with ambiguity. Some believe that the nurarihyon was not actually the leader of the Night Parade, but that this was based on a misunderstanding/mistranslation. For further information in English, see Michael Dylan Foster, *The Book of Yōkai: Mysterious Creatures of Japanese Folklore* (Oakland: University of California Press, 2015), 218.
13. Kenji, "Hyakki Yagyō," in *Yokai Jiten*.

CHAPTER 10: THE HOUR OF THE OX

1. David Luke and Karolina Zychowicz, "Working the Graveyard Shift at the Witching Hour: Further Exploration of Dreams, Psi and Circadian Rhythms," *International Journal of Dream Research* 7, no. 4 (2014), 109, https://gala.gre.ac.uk/id/eprint/12514/1/2014_-_IJDR_-_Luke_&_Zychowicz.pdf.
2. Komatsu Kazuhiko, *An Introduction to Yōkai Culture*, trans. Hiroko Yoda and Matt Alt (Tokyo: Japan Publishing Industry Foundation for Culture, 2017), 137.
3. Yanagita Kunio's theories of yūrei are outlined in Michael Dylan Foster, *Pandemonium and Parade* (Berkeley: University of California Press, 2009), 152–54.
4. Ikeda Yasaburō, *Nihon no Yūrei* (Tokyo: Chūō Kōron Shinsha, 2002), trans. Lisa Hofmann-Kuroda.
5. Komatsu, *Yōkai Culture*, 148. Yasunaga Toshinobu lived from 1929 to 1995.
6. Komatsu, "What Is a Yōkai," trans. Kaya Laterman and Satori Murata, in *Yōkai: Ghosts, Demons and Monsters of Japan*, ed. Felicia Katz-Harris (Santa Fe: Museum of New Mexico Press, 2019), 149.

PART 4: KETSU

1. Imai Takajirō, *Sakubun no Jugyō Nyūmon* (Tokyo: Meiji Tosho Shuppan, 1961), 187, trans. Lisa Hofmann-Kuroda.
2. Suzuki Toshio, "Yōsai no 'Kishōtenketsu' Sesshakujun," *Tōyō Shihō* 3 (1997): 8–18, trans. Lisa Hofmann-Kuroda.

CHAPTER 11: MOURNINGTIME

1. Zack Davisson, "Baku—The Dream Eater," *Hyakumonogatari Kaidankai*, October 20, 2012, https://hyakumonogatari.com/2012/10/20/baku-the-dream-eater/. Davisson gathers and translates his baku information from various Japanese sources including Mizuki Shigeru's *Mujara*.

2.   Erika Kaneko, "The Death Ritual," in *Ryukyuan Culture and Society: A Survey*, ed. Allan H. Smith (Honolulu: University of Hawaii Press, 1964), 25–29.

3.   Ibid.

4.   Ibid.

5.   Kao Kalia Yang and Shannon Gibney, eds., *What God Is Honored Here?: Writings on Miscarriage and Infant Loss by and for Native Women and Women of Color* (Minneapolis: University of Minnesota Press, 2019).

6.   Julian Baumgartner, *Scraping, Scraping, Scraping, or a Slow Descent into Madness*, YouTube video, 40:59, posted by "Baumgartner Restoration," November 11, 2019, https://www.youtube.com/watch?v=YOOQl0hC18U.

7.   Christopher T. Nelson, *Dancing with the Dead: Memory, Performance, and Everyday Life in Postwar Okinawa* (Durham, NC: Duke University Press, 2008), 147–48.

8.   Kaneko, "The Death Ritual," 25–29.

9.   Ibid.

10. Ibid. Some scholars debate whether this third stage is actually distinct from the second stage.

11. "Mo Is a Monster That Consumes Dreams, They Eat Away One's Nightmares," [每日頭條] *Mei Ri Tou Tiao*, June 29, 2017, https://kknews.cc/zh-tw/history/68y459l.html, trans. Jenna Tang.

12. *Nihon Mikakunin Seibutsu Jiten*, ed. Sasama Yoshihiko (Tokyo: Kadokawa, 2018), Kindle, trans. Lisa Hofmann-Kuroda.

13. Song of Solomon 5:2a.

CHAPTER 12: THE NAMING

1.   "Important Milestones: Your Child by One Year," Centers for Disease Control and Prevention, revised February 7, 2022, https://www.cdc.gov/ncbddd/actearly/milestones/milestones-1yr.html.

2.   Michael Dylan Foster, *The Book of Yōkai: Mysterious Creatures of Japanese Folklore* (Oakland: University of California Press, 2015), 107.

3.   This tale is transformed from sources including "Amabie no shōtai wo otte: Sugata mita mono, shi wo nogarerareru amabiko no hakken," *Mainichi shimbun sha*, June 5, 2020, https://mainichi.jp/articles/20200605/ddl/k18/040/220000c, trans. Lisa Hofmann-Kuroda.

4.   Rebecca Saunders, "Amabie: The Japanese Monster Going Viral," BBC Travel, April 23, 2020, https://www.bbc.com/travel/article/20200422-amabie-the-japanese-monster-going-viral.

5.   The last letter "ko" in アマビコ (a-ma-bi-ko) looked similar to the old letter "ye" (which no longer exists in modern Japanese), so it was easy to misread "ko" for "ye," Lisa Hofmann-Kuroda explains. See "Amabie naranu Amabiko Ashikaga Gakkō Kobunsho ni sanbon ashi no yōkai," *Tokyo shimbun*, May 29, 2020, https://www.tokyo-np.co.jp/article/31969.

6.   Amache, originally named the Granada War Relocation Authority Center, was based in Colorado, and held over ten thousand Japanese Americans between 1942 and 1945.

7.   Takada Tomoki, "Chiiki Hōsai to Yōkai," *Gakujutsu no Dōkō* 25, no. 11 (2020): 44–48, trans. Lisa Hofmann-Kuroda.

CHAPTER 13: THE YEAR OF THE RAT

1.   Fercility Jiang, "Rat Chinese Zodiac Sign: Symbolism in Chinese Culture," *China Highlights*, revised January 10, 2023, https://www.chinahighlights.com/travelguide/chinese-zodiac/rat-chinese-zodiac-sign-symbolism.htm.

2.   This story is transformed from Pu Songling's tale, written between the 1670s and the 1700s (during the Qing dynasty), and first published in 1766. Translated and recounted by Jenna Tang, the story is available in English in Pu's *Strange Tales from a Chinese Studio* (New York: Penguin Classics, 2006).

3.   "The Constellations," International Astronomical Union, October 13, 2022, https://www.iau.org/public/themes/constellations/.

4.   "88 Officially Recognized Constellations," High Energy Astrophysics Science Archive Research Center, October 13, 2022, https://starchild.gsfc.nasa.gov/docs/StarChild/questions/88constellations.html.

5.   "The Constellations," International Astronomical Union.

6.   "The Big Dipper Is Not a Constellation. So What Is It?" *Look at the Sky*, May 11, 2020, https://lookatthesky.com/big-dipper-is-not-a-constellation/.

7.   Junjun Xu, "Ancient Chinese Constellations," *Proceedings of the International Astronomical Union* 5, no. S260 (2009): 107–15, https://doi.org/10.1017/S174392131100319X.

8.   Wilt L. Idema, "Old Tales for New Times: Some Comments on the Cultural Translation of China's Four Great Folktales in the Twentieth Century," *Taiwan Journal of East Asian Studies* 9, no. 1 (2012).

9.   "The Constellations," International Astronomical Union.

10. "Event Horizon," Wikipedia, last edited February 23, 2023, https://en.wikipedia.org/wiki/Event_horizon.

11. Dennis Overbye, "The Milky Way's Black Hole Comes to Light," *New York Times*, May 12, 2022, https://www.nytimes.com/2022/05/12/science/black-hole-photo.html.

12. Her ideas are also expressed in her article: Yvette Cendes, "How Do Black Holes Swallow Stars?" *Astronomy*, December 8, 2021, https://astronomy.com/magazine/news/2021/12/how-do-black-holes-swallow-stars.

CHAPTER 14: IN THE WHIRLPOOLS

1. Nahoko Kahara, "From Folktale Hero to Local Symbol: The Transformation of Momotaro (the Peach Boy) in the Creation of a Local Culture," *Waseda Journal of Asian Studies* 25 (2004): 35–62.
2. For more about the history of Ainu (and particularly, Ainu in diaspora) I recommend the work of Dr. R. Māpuana Shizuko Hayashi-Simpliciano, particularly the film *Ainu in Diaspora*, https://vimeo.com/538992254.
3. For more, see the chapter "Oni and Japanese Identity" in Noriko T. Reider, *Japanese Demon Lore* (Logan: Utah State University Press, 2010).
4. Quoted from Iwaya Sazanami's popular *Nihon mukashibanashi* children's fairy-tale series, published between 1894 and 1896, as translated in Reider, *Japanese Demon Lore*, 107.
5. Reider, *Japanese Demon Lore*, 107.
6. Ibid.
7. Quotes taken from my grandfather Tom Nakamura's autobiographical school paper that he completed in 1944 while in Amache. He was sixteen.
8. From my grandfather Tom Nakamura's current memoir project, *A Time in History*. As of publication, he is ninety-four years old and still diligently working on it in Google Docs.
9. Caps, exclamation points, elisions, and emphasis mine; text excerpted from Andrew Yang, "We Asians Are Not the Virus, But We Can Be Part of the Cure," *Washington Post*, April 1, 2020, https://www.washingtonpost.com/opinions/2020/04/01/andrew-yang-coronavirus-discrimination/.
10. From the work of Takahashi Masaaki, as explained in Reider, *Japanese Demon Lore*, 49.
11. Steven C. Ridgely, "Terayama Shūji and Bluebeard," *Marvels & Tales* 27, no. 2 (2013), https://digitalcommons.wayne.edu/marvels/vol27/iss2/10.

CHAPTER 15: AUTOMYTHOLOGIES II (REPRISE)

1. Toriyama's set of yōkai encyclopedias are translated and annotated in Toriyama Sekien, *Japandemonium Illustrated: The Yokai Encyclopedias of Toriyama Sekien*, trans. Hiroko Yoda and Matt Alt (New York: Dover, 2017), which is where I encountered them.
2. Noriko T. Reider, "Animating Objects: Tsukumogami ki and the Medieval Illustration of Shingon Truth," *Japanese Journal of Religious Studies* 36, no. 2 (2009): 231–57.
3. Toriyama, *Japandemonium*, viii.
4. Angie Chuang, "Scars, Silence, and Diane Fossey," in *A Harp in the Stars: An Anthology of Lyric Essays*, ed. Randon Billings Noble (Lincoln: University of Nebraska Press, 2021), 71.
5. Michiko Iwasaka and Barre Toelken, *Ghosts and the Japanese* (Logan: Utah State University Press, 1994), 6.
6. Leslie Jamison, ed., "Introduction," in *Best American Essays 2017* (Boston: Houghton Mifflin, 2017), Kindle.
7. Clarice Lispector, *The Hour of the Star*, trans. Giovanni Pontiero (New York: New Directions, 1992), Kindle.

CHAPTER 16: THE THREE CORPSES

1. *Taishang linbao wufuxu*. This ancient text is from the Daozang, the Daoist Canon, and is uploaded at Chinese Text Project, https://ctext.org/wiki.pl?if=gb&res=993764, trans. Jenna Tang.
2. Gil Raz, "Imbibing the Universe: Methods of Ingesting the Five Sprouts," *Asian Medicine* 7, no. 1 (2012): 65–100, https://doi.org/10.1163/15734218-12341244.
3. *Taishang Chu San Shi Jiu Chong Bao Sheng Jing*. This text was also collected from the Daozang, the Daoist Canon, and archived at https://ctext.org/wiki.pl?if=gb&chapter=333930, trans. Jenna Tang.
4. Shailesh M. Gondivkar, Amol Gadbail, and Revant Chole, "Oral Pregnancy Tumor," *Contemporary Clinical Dentistry* 1, no. 3 (2010): 190–92, https://www.ncbi.nlm.nih.gov/pmc/articles/PMC3220110/.
5. From the classical *T'ai-shang sau-shih chung-ching* as quoted in Henry Maspero, *Taoism and Chinese Religion*, trans. Frank A. Kierman Jr. (Amherst: University of Massachusetts Press, 1981), 332.
6. Raz, "Imbibing the Universe."
7. Kenkō, "Essays on Idleness," trans. Donald Keene, in *Traditional Japanese Literature: Traditional Japanese Literature: An Anthology, Beginnings to 1600*, ed. Haruo Shirane (New York: Columbia University Press, 2007), 828.
8. Michael Dylan Foster, *The Book of Yōkai: Mysterious Creatures of Japanese Folklore* (Oakland: University of California Press, 2015), 19.
9. Michael Dylan Foster, *Pandemonium and Parade* (Berkeley: University of California Press, 2009), 10.
10. Lucia Berlin, "Silence," in *A Manual for Cleaning Women: Selected Stories* (New York: Picador, 2016), 3211.
11. Michiko Iwasaka and Barre Toelken, *Ghosts and the Japanese* (Logan: Utah State University Press, 1994), 57.

BIBLIOGRAPHY

Acocella, Joan. "Once Upon a Time: The Lure of the Fairy Tale." *The New Yorker*, July 16, 2012. https://www
.newyorker.com/magazine/2012/07/23/once-upon-a-time-3.

"Amabie naranu Amabiko Ashikaga Gakkō Kobunsho ni sanbon ashi no yōkai." *Tokyo shimbun*, May 29, 2020.
https://www.tokyo-np.co.jp/article/31969.

"Amabie no shōtai wo otte: Sugata mita mono, shi wo nogarerareru amabiko no hakken." *Mainichi shimbun sha*,
June 5, 2020. https://mainichi.jp/articles/20200605/ddl/k18/040/220000c.

Bathgate, Michael. *The Fox's Craft in Japanese Religion and Culture: Shapeshifters, Transformations, and Duplicities.*
New York: Routledge, 2004.

Benjamin, Walter. *The Storyteller Essays.* Translated by Tess Lewis. New York: NYRB Classics, 2019. Kindle.

Berlin, Lucia. *A Manual for Cleaning Women: Selected Stories.* New York: Picador, 2016.

"The Big Dipper Is Not a Constellation. So What Is It?" *Look at the Sky*, May 11, 2020. https://lookatthesky
.com/big-dipper-is-not-a-constellation/.

Bloom, Dan. "'There She Blows!'" *Daily Mail*, November 27, 2013. https://www.dailymail.co.uk/news
/article-2514317/Horrifying-footage-shows-washed-sperm-whale-EXPLODING-biologist-tries-cut-carcass
.html.

Brina, Elizabeth Miki. *Speak, Okinawa: A Memoir.* New York: Knopf, 2021. Kindle.

Burgess, Lana. "What Is Psychomotor Agitation?" Medical News Today, Healthline Media, March 9, 2022.
https://www.medicalnewstoday.com/articles/319711.

Cendes, Yvette. "How Do Black Holes Swallow Stars?" *Astronomy*, December 8, 2021. https://astronomy.com
/magazine/news/2021/12/how-do-black-holes-swallow-stars.

Chou Yu-Lan, Ding Yi-Tsai, and the Committee of Chinese Classical Literature. *Bei She Chuan Tong Su Ben*
[The Popular White Snake Legend]. Taipei: Silkbook, 2007.

Chuang, Angie. "Scars, Silence, and Diane Fossey." In *A Harp in the Stars: An Anthology of Lyric Essays*, edited by
Randon Billings Noble. Lincoln: University of Nebraska Press, 2021.

Cirlot, Juan Eduardo. *A Dictionary of Symbols.* Translated by Jack Sage and Valerie Miles. New York: NYRB
Books, 2020.

Davis, F. Hadland. *Myths and Legends of Japan.* Mineola, NY: Dover, [1913] 1992.

Davisson, Zack. "Bakekujira and Japan's Whale Cults." *Hyakumonogatari Kaidankai*, May 10, 2015. https://hyakumonogatari.com/2013/05/10/bakekujira-and-japans-whale-cults/.

———. "Baku—The Dream Eater." *Hyakumonogatari Kaidankai*, October 20, 2012. https://hyakumonogatari.com/2012/10/20/baku-the-dream-eater/.

———. "Kitsune no Yomeiri–The Fox Wedding." *Hyakumonogatari Kaidankai*, July 19, 2013. https://hyakumonogatari.com/2013/07/19/kitsune-no-yomeiri-the-fox-wedding/.

———. "Introduction." In *Tono Monogatari*, by Mizuki Shigeru, translated by Zack Davisson. Montreal: Drawn & Quarterly, 2021.

De Visser, Marinus Willem. *The Dragon in China and Japan*. Amsterdam: J. Müller, 1913.

Dickens, Charles. *A Tale of Two Cities*. New York: Bantam Classics, 2003.

Ferrari, Thomas E. "Cetacean Beachings Correlate with Geomagnetic Disturbances in Earth's Magnetosphere: An Example of How Astronomical Changes Impact the Future of Life." *International Journal of Astrobiology* 16, no. 2 (2017): 163–75. https://doi.org/10.1017/S1473550416000252.

Figal, Gerald. *Civilization and Monsters: Spirits of Modernity in Meiji Japan*. Durham, NC: Duke University Press, 2000.

Foster, Michael Dylan. *The Book of Yokai: Mysterious Creatures of Japanese Folklore*. Oakland: University of California Press, 2015.

———. "The Metamorphosis of the Kappa: Transformation of Folklore to Folklorism in Japan." *Asian Folklore Studies* 57, no. 1 (1998): 1–24.

———. *Pandemonium and Parade*. Berkeley: University of California Press, 2009.

Gondivkar, Shailesh M., Amol Gadbail, and Revant Chole. "Oral Pregnancy Tumor." *Contemporary Clinical Dentistry* 1, no. 3 (2010): 190–92. https://www.ncbi.nlm.nih.gov/pmc/articles/PMC3220110/.

Grapard, Allan G. "The Shinto of Yoshida Kanetomo." *Monumenta Nipponica* 47, no. 1 (1992): 27–58. https://doi.org/10.2307/2385357.

Grimm, Bruce Owens. "Haunted Memoir." *Assay: A Journal of Nonfiction Studies* 7, no. 1 (2020). https://www.assayjournal.com/bruce-owens-grimm-haunted-memoir-assay-71.html.

Hearn, Lafcadio. *Glimpses of Unfamiliar Japan*. Tokyo: Tuttle Publishing, [1894] 2009.

———. *Kwaidan: Stories and Studies of Strange Things*. North Clarendon, VT: Tuttle Publishing, [1904] 1971.

Hibbett, Maia. "In Okinawa, the US Military Seeks a Base Built on the Bones of the War Dead." *The Nation*, February 18, 2021. https://www.thenation.com/article/world/japan-okinawa-henoko/.

Hino Iwao. *Dōbutsu Yōkai Tan, Volume 1*. Tokyo: Chuko Koron Shinsha, [1926] 2006.

Idema, Wilt. "Old Tales for New Times: Some Comments on the Cultural Translation of China's Four Great Folktales in the Twentieth Century." *Taiwan Journal of East Asian Studies* 9, no. 1 (2012): 1–23.

Ikeda Yasaburō. *Nihon no Yūrei*. Tokyo: Chūō Kōron Shinsha, 2002.

Imai Takajirō. *Sakubun no Jugyō Nyūmon*. Tokyo: Meiji Tosho Shuppan, 1961.

"Important Milestones: Your Child by One Year." Centers for Disease Control and Prevention, revised February 7, 2022. https://www.cdc.gov/ncbddd/actearly/milestones/milestones-1yr.html.

Itoh, Mayumi. *The Japanese Culture of Mourning Whales: Whale Graves and Memorial Monuments in Japan.* London: Palgrave Macmillan, 2018.

Ivy, Marilyn. *Discourses of the Vanishing: Modernity, Phantasm, Japan.* Chicago: University of Chicago Press, 1995.

Iwasaka, Michiko, and Barre Toelken. *Ghosts and the Japanese: Cultural Experience in Death Legends.* Logan: Utah State University Press, 1994.

Jamison, Leslie, ed. *Best American Essays 2017.* Boston: Houghton Mifflin, 2017. Kindle.

Jiang, Fercility. "Rat Chinese Zodiac Sign: Symbolism in Chinese Culture." *China Highlights,* revised January 10, 2023. https://www.chinahighlights.com/travelguide/chinese-zodiac/rat-chinese-zodiac-sign-symbolism.htm.

Kaneko, Erika. "The Death Ritual." In *Ryukyuan Culture and Society: A Survey,* edited by Allan H. Smith, 25–29. Honolulu: University of Hawaii Press, 1964.

Keisei. "A Companion in Solitude." Translated by Michael Emmerich. In *Traditional Japanese Literature: An Anthology, Beginnings to 1600,* edited by Haruo Shirane, 292–94. New York: Columbia University Press, 2007.

Kenkō. "Essays on Idleness." Translated by Donald Keene. In *Traditional Japanese Literature: An Anthology, Beginnings to 1600,* edited by Haruo Shirane, 820–43. New York: Columbia University Press, 2007.

Kim, H. Yumi. *Madness in the Family: Women, Care, and Illness in Japan.* New York: Oxford University Press, 2022.

The Kojiki: Records of Ancient Matters. Translated by Basil Hall Chamberlain. North Clarendon, VT: Tuttle Publishing, [1882] 1981.

Komatsu Kazuhiko. *An Introduction to Yōkai Culture.* Translated by Hiroko Yoda and Matt Alt. Tokyo: Japan Publishing Industry Foundation for Culture, 2017.

———. "What Is a Yōkai." Translated by Kaya Laterman and Satori Murata. In *Yokai: Ghosts, Demons and Monsters of Japan,* edited by Felicia Katz-Harris. Santa Fe: Museum of New Mexico Press, 2019.

Liao Hung-chi. "Stranded." Translated by Jacqueline Li. *The Willowherb Review,* December 2021. https://www.thewillowherbreview.com/stranded-liao-hungchi-.

Lispector, Clarice. *The Hour of the Star.* Translated by Giovanni Pontiero. New York: New Directions, 1992. Kindle.

Luke, David, and Karolina Zychowicz. "Working the Graveyard Shift at the Witching Hour: Further Exploration of Dreams, Psi and Circadian Rhythms." *International Journal of Dream Research* 7, no. 4 (2014): 105–12. https://gala.gre.ac.uk/id/eprint/12514/1/2014_-_IJDR_-_Luke_&_Zychowicz.pdf.

Machado, Carmen Maria. *In the Dream House: A Memoir.* Minneapolis: Graywolf Press, 2019.

Maspero, Henry. *Taoism and Chinese Religion.* Translated by Frank A. Kierman Jr. Amherst: University of Massachusetts Press, 1981.

Matsuzaki Kenzō. "Yorikujira no shochi wo megutte: Dōshokubutsu no kuyō." *Nihon Jōmin bunka kiyō* 19 (1996).

Meyer, Matthew. *The Book of the Hakutaku: A Bestiary of Japanese Monsters.* Self-published, 2018.

———. *The Hour of Meeting Evil Spirits: An Encyclopedia of Mononoke and Magic.* Self-published, 2015.

———. "Kyubi no Kitsune." *A Yōkai a Day,* October 7, 2007. https://matthewmeyer.net/blog/2009/10/07/a-yōkai-a-day-kyubi-no-kitsune.

————. *The Night Parade of One Hundred Demons: A Field Guide to Japanese Yōkai.* Self-published, 2015.

Mizuki Shigeru. *Mizuki Shigeru no Sekai Genjū Jiten.* Tokyo: Asahi Shinbunsha, 1994.

"Mo Is a Monster That Consumes Dreams, They Eat Away One's Nightmares." *Mei Ri Tou Tiao,* June 29, 2017. https://kknews.cc/zh-tw/history/68y459l.html.

Murai, Mayako. *From Dog Bridegroom to Wolf Girl: Contemporary Japanese Fairy-Tale Adaptations in Conversation with the West.* Detroit: Wayne State University Press, 2015.

Murakami Kenji. *Yōkai Jiten.* Tokyo: Mainichi Shimbunsha, 2013. Kindle.

Nahoko Kahara. "From Folktale Hero to Local Symbol: The Transformation of Momotaro (the Peach Boy) in the Creation of a Local Culture." *Waseda Journal of Asian Studies* 25 (2004): 35–62.

Nelson, Christopher T. *Dancing with the Dead: Memory, Performance, and Everyday Life in Postwar Okinawa.* Durham, NC: Duke University Press, 2008.

Ōe Osamu, ed. *Tamashī no Minzokugaku: Tanigawa Kenichi no sekai.* Tokyo: Tomiyamabō Intanashonaru, 2006.

Overbye, Dennis. "The Milky Way's Black Hole Comes to Light." *New York Times,* May 12, 2022. https://www.nytimes.com/2022/05/12/science/black-hole-photo.html.

Ozaki, Yei Theodora. *The Japanese Fairy Book.* New York: Dover, [1903] 1967.

Pandey, Rajyashree. "Women, Sexuality, and Enlightenment: Kankyo No Tomo." *Monumenta Nipponica* 50, no. 3 (1995): 325–56. https://doi.org/10.2307/2385548.

Pham, Larissa. *Pop Song: Adventures in Art and Intimacy.* New York: Catapult, 2021.

Pickle, Betsy. "Al-Anon Helps Family, Friends to Orderly Lives." *Knoxville News-Sentinel,* October 11, 1981.

Pu Songling. *Strange Tales from a Chinese Studio.* New York: Penguin Classics, 2006.

Raz, Gil. "Imbibing the Universe: Methods of Ingesting the Five Sprouts." *Asian Medicine* 7, no. 1 (2012): 65–100. https://doi.org/10.1163/15734218-12341244.

Reider, Noriko T. "Animating Objects: Tsukumogami ki and the Medieval Illustration of Shingon Truth." *Japanese Journal of Religious Studies* 36, no. 2 (2009): 231–57.

————. "The Appeal of 'Kaidan,' Tales of the Strange." *Asian Folklore Studies* 59, no. 2 (2000): 265–83. https://doi.org/10.2307/1178918.

————. *Japanese Demon Lore.* Logan: Utah State University Press, 2010.

————. *Seven Demon Stories from Medieval Japan.* Boulder: University Press of Colorado, 2016.

Ridgely, Steven C. "Terayama Shūji and Bluebeard." *Marvels & Tales* 27, no. 2 (2013): 290–300. https://www.jstor.org/stable/10.13110/marvelstales.27.2.0290.

Sakade, Florence. *Japanese Children's Favorite Stories.* North Clarendon, VT: Tuttle Publishing, 2014.

————. *Urashima Tarō and Other Japanese Children's Stories.* Rutland, VT: Tuttle Publishing, 1986.

Salesses, Matthew. *Craft in the Real World: Rethinking Fiction Writing and Workshopping.* New York: Catapult, 2021.

Sasama Yoshihiko, ed. *Nihon Mikakunin Seibutsu Jiten.* Tokyo: Kadokawa, 2018. Kindle.

Saunders, Rebecca. "Amabie: The Japanese Monster Going Viral." *BBC Travel*, April 23, 2020. https://www.bbc.com/travel/article/20200422-amabie-the-japanese-monster-going-viral.

Shibata Shōkyoku, ed. *Kidan Ibun Jiten*. Tokyo: Chikuma Shobō, 2008.

Shimabukuro Genshichi. "Okinawa ni okeru yorimono." *Minzoku Denshō* 15, no. 11 (1951).

Shūji Tomita. *Ehagaki de Miru Nihon Kindai*. Tokyo: Seikyusha, 2005.

Sontag, Susan. *Illness as Metaphor and AIDS and Its Metaphors*. New York: Anchor, 1990.

Suzuki Toshio. "Yōsai no 'Kishōtenketsu' Sesshakujun." *Tōyō Shihō* 3 (1997): 8–18.

Tada Katsumi, and Kyōgoku Natsuhiko. *Yōkai Zukan*. Tokyo: Kokusho Kankōkai, 2000.

Takada Tomoki. "Chiiki Hōsai to Yōkai," *Gakujutsu no Dōkō* 25, no. 11 (2020): 44–48.

Tamir, Diana I., et al. "Media Usage Diminishes Memory for Experiences." *Journal of Experimental Social Psychology* 76 (2018): 161–68.

Tanji, Miyume. *Myth, Protest and Struggle in Okinawa*. New York: Routledge, 2006.

Taylor, Alan. "5 Years Since the 2011 Great East Japan Earthquake." *The Atlantic*, March 10, 2016. https://www.theatlantic.com/photo/2016/03/5-years-since-the-2011-great-east-japan-earthquake/473211.

Toriyama Sekien. *Japandemonium Illustrated: The Yokai Encyclopedias of Toriyama Sekien*. Translated by Hiroko Yoda and Matt Alt. New York: Dover, 2017.

Uchida Senzō. "Basic Information Regarding a Rare and Wild Aquatic Creature of Japan." *Ministry of Agriculture, Forestry and Fishing: Fisheries Agency Volume*. Japanese Fisheries Agency, 1994.

Uesedo Tōru. *Taketomi Tōshi Minwa*. Tokyo: Hōsei Daigaku Shuppan Kyoku, 1976.

United Nations Environment Programme. *Dugong Status Report and Action Plans for Countries and Territories*. Edited by Helene Marsh. 2002. https://portals.iucn.org/library/sites/library/files/documents/2002-001.pdf.

Van der Kolk, Bessel A. *The Body Keeps the Score: Brain, Mind, and Body in the Healing of Trauma*. New York: Penguin Books, 2015. Kindle.

Vieta, Eduard, et al., "Protocol for the Management of Psychiatric Patients with Psychomotor Agitation." *BMC Psychiatry* 17, no. 1 (August 2017). https://doi.org/10.1186/s12888-017-1490-0.

"Whale Explodes in a Taiwanese City." BBC News, January 29, 2004. http://news.bbc.co.uk/2/hi/science/nature/3437455.stm.

"With a Population Under a Dozen, Okinawa Dugongs Labeled 'Critically Endangered' by IUCN." *The Japan Times*, December 12, 2019. https://www.japantimes.co.jp/news/2019/12/12/national/science-health/okinawa-dugongs-critically-endangered.

Xu, Junjun. "Ancient Chinese Constellations." *Proceedings of the International Astronomical Union* 5, no. S260 (2009): 107–15. https://doi.org/10.1017/S1743921311000319X.

Yanagihara, Hanya. *A Little Life*. New York: Doubleday, 2015.

Yanagita Kunio and Sasaki Kizen. *Folk Legends from Tōno: Japan's Spirits, Deities, and Phantastic Creatures*. Translated by Ronald Morse. Lanham, MD: Rowman & Littlefield, 2015.

———. *The Yanagita Kunio Guide to the Japanese Folktale*. Translated by Fanny Hagin Mayer. Bloomington: Indiana University Press, 1986.

Yang, Andrew. "We Asians Are Not the Virus, But We Can Be Part of the Cure." *Washington Post*, April 1, 2020. https://www.washingtonpost.com/opinions/2020/04/01/andrew-yang-coronavirus-discrimination/.

Yang, Kao Kalia, and Shannon Gibney, eds. *What God Is Honored Here?: Writings on Miscarriage and Infant Loss by and for Native Women and Women of Color*. Minneapolis: University of Minnesota Press, 2019.

Yoda, Hiroko, and Matt Alt. *Yōkai Attack!* North Clarendon, VT: Tuttle Publishing, 2012.

Yonamine, Kaiya. *Our Island's Treasure*. https://vimeo.com/340517922.

Yun Ying Tian Guang. "Brief Analysis of Different Versions of White Snake Legend (The Origin of the White Snake): A Review." *Weiwenku*, January 20, 2019. https://www.gushiciku.cn/dc_dr/KGV6.

About the Author

JAMI NAKAMURA LIN is a Japanese Taiwanese Okinawan American writer, whose work has been featured in the *New York Times*, Catapult, and *Electric Literature*, among other publications. She has received fellowships and support from the National Endowment for the Arts/Japan-U.S. Friendship Commission, Yaddo, Sewanee Writers' Conference, We Need Diverse Books, and the Illinois Arts Council. She received her MFA in nonfiction from Pennsylvania State University and lives in the Chicago area with her husband and family.

About the Illustrator

CORI NAKAMURA LIN is a Japanese Taiwanese American illustrator specializing in culture-centered radical storytelling. She is the illustrator of *When Everything Was Everything* by Saymoukda Duangphouxay Vongsay. Her work has been featured in the *Los Angeles Times* and Eater Chicago as well as on the History Channel, PBS's Michigan Learning Channel, and WBEZ Chicago. She lives in Chicago.

HarperCollins books may be purchased for educational, business, or sales
promotional use. For information, please email the Special Markets Department
at SPsales@harpercollins.com.

FIRST EDITION

Designed by Leah Carlson-Stanisic

Library of Congress Cataloging-in-Publication Data has been applied for.

ISBN 978-0-06-321323-4

23 24 25 26   RTL   10 9 8 7 6 5 4 3 2 1